Immunodiagnostics

The Practical Approach Series

SERIES EDITOR

B. D. HAMES
Department of Biochemistry and Molecular Biology
University of Leeds, Leeds LS2 9JT, UK

See also the Practical Approach web site at **http://www.oup.co.uk/PAS**

★ indicates new and forthcoming titles

Affinity Chromatography
Affinity Separations
Anaerobic Microbiology
Animal Cell Culture (2nd edition)
Animal Virus Pathogenesis
Antibodies I and II
Antibody Engineering
★ Antisense Technology
Applied Microbial Physiology
Basic Cell Culture
Behavioural Neuroscience
Bioenergetics
Biological Data Analysis
Biomechanics—Materials
Biomechanics—Structures and Systems
Biosensors
Carbohydrate Analysis (2nd edition)
Cell-Cell Interactions
The Cell Cycle
Cell Growth and Apoptosis
★ Cell Separation
Cellular Calcium
Cellular Interactions in Development
Cellular Neurobiology
★ Chromatin
★ Chromosome Structural Analysis
Clinical Immunology
Complement
★ Crystallization of Nucleic Acids and Proteins (2nd edition)
Cytokines (2nd edition)
The Cytoskeleton
Diagnostic Molecular Pathology I and II
DNA and Protein Sequence Analysis
DNA Cloning 1: Core Techniques (2nd edition)
DNA Cloning 2: Expression Systems (2nd edition)
DNA Cloning 3: Complex Genomes (2nd edition)
DNA Cloning 4: Mammalian Systems (2nd edition)

★ Drosophila (2nd edition)
Electron Microscopy in Biology
Electron Microscopy in Molecular Biology
Electrophysiology
Enzyme Assays
Epithelial Cell Culture
Essential Developmental Biology
Essential Molecular Biology I and II
★ Eukaryotic DNA Replication
Experimental Neuroanatomy
Extracellular Matrix
Flow Cytometry (2nd edition)
Free Radicals
Gas Chromatography
Gel Electrophoresis of Nucleic Acids (2nd edition)
★ Gel Electrophoresis of Proteins (3rd edition)
Gene Probes 1 and 2
Gene Targeting
Gene Transcription
★ Genome Mapping
Glycobiology
★ Growth Factors and Receptors
Haemopoiesis
★ High Resolution Chromatography
Histocompatibility Testing
HIV Volumes 1 and 2
★ HPLC of Macromolecules (2nd edition)
Human Cytogenetics I and II (2nd edition)
Human Genetic Disease Analysis
★ Immobilized Biomolecules in Analysis
Immunochemistry 1
Immunochemistry 2
Immunocytochemistry
★ *In Situ* Hybridization (2nd edition)
Iodinated Density Gradient Media
Ion Channels
★ Light Microscopy (2nd edition)
Lipid Modification of Proteins
Lipoprotein Analysis
Liposomes
Mammalian Cell Biotechnology
Medical Parasitology
Medical Virology
MHC Volumes 1 and 2
★ Molecular Genetic Analysis of Populations (2nd edition)
Molecular Genetics of Yeast
Molecular Imaging in Neuroscience
Molecular Neurobiology
Molecular Plant Pathology I and II
Molecular Virology
Monitoring Neuronal Activity
Mutagenicity Testing
★ Mutation Detection
Neural Cell Culture
Neural Transplantation
Neurochemistry (2nd edition)

Neuronal Cell Lines
NMR of Biological Macromolecules
Non-isotopic Methods in Molecular Biology
Nucleic Acid Hybridization
Oligonucleotides and Analogues
Oligonucleotide Synthesis
PCR 1
PCR 2
★ PCR 3: PCR *In Situ* Hybridization
Peptide Antigens
Photosynthesis: Energy Transduction
Plant Cell Biology
Plant Cell Culture (2nd edition)
Plant Molecular Biology
Plasmids (2nd edition)
Platelets
Postimplantation Mammalian Embryos
★ Post-Translational Modification Preparative Centrifugation
Protein Blotting
★ Protein Expression
Protein Engineering
Protein Function (2nd edition)
Protein Phosphorylation
Protein Purification Applications
Protein Purification Methods
Protein Sequencing
Protein Structure (2nd edition)
Protein Structure Prediction
Protein Targeting
Proteolytic Enzymes
Pulsed Field Gel Electrophoresis
RNA Processing I and II
★ RNA-Protein Interactions
Signalling by Inositides
Subcellular Fractionation
Signal Transduction
★ Transcription Factors (2nd edition)
Tumour Immunobiology

Immunodiagnostics

A Practical Approach

Edited by

RAYMOND EDWARDS
Director of Netria,
St. Bartholomew's Hospital, London

OXFORD
UNIVERSITY PRESS
Great Clarendon Street, Oxford OX2 6DP
Oxford University Press is a department of the University of Oxford
and furthers the University's aim of excellence in research, scholarship,
and education by publishing worldwide in
Oxford New York
Athens Auckland Bangkok Bogotá Buenos Aires Calcutta
Cape Town Chennai Dar es Salaam Delhi Florence Hong Kong Istanbul
Karachi Kuala Lumpur Madrid Melbourne Mexico City Mumbai
Nairobi Paris São Paulo Singapore Taipei Tokyo Toronto Warsaw
and associated companies in Berlin Ibadan
Oxford is a registered trade mark of Oxford University Press
Published in the United States
by Oxford University Press Inc., New York

A catalogue record for this book is available from the British Library
Library of Congress Cataloging in Publication Data
(Data available)
ISBN 0-19-963589-7 (Hbk)
0-19-963588-9 (Pbk)
Typeset by Footnote Graphics,
Warminster, Wilts
Printed in Great Britain by Information Press, Ltd,
Eynsham, Oxon.

Preface

It has been four decades since the publication of the principles of radioimmunoassay. These studies, based on work in London by Ekins and in New York by Berson and Yalow, must be considered as milestones in the development of modern analytical methods. They have had a major and decisive influence on the development of diagnostic tests. The term 'immunodiagnostics' used in the title means diagnostic tests based on the principles of immunoassay, where the primary reagent is an antibody.

The widespread dissemination of these tests, particularly in the lucrative clinical field, has been accompanied by considerable commercial involvement. Undoubtedly this proliferation has made many methods accessible and easy to use. However, competition has also led to constraints on the availability of technical and scientific information. As a consequence, this has introduced limitations for flexibility or interactive use. The problems are compounded when methods are integrated into 'black box' instrumentation, as are often found in many types of automated analysers.

Compiling an encyclopaedic publication covering all aspects of immunodiagnostic methodology would be well beyond the scope of a single volume and certainly irrelevant for the 'Practical Approach Series'. The primary focus in this book has been the analytical format itself; the primary reagent (i.e. the antibody molecule) in its various forms (e.g. solid phase; purified molecule; fragment; etc.); particular types of signal detection; and the fundamental aspects applicable to all tests (i.e. optimisation and validation). The aim has been that readers should be able to use any of the methods for their own particular application but also to appreciate the underlying principles involved in the choice of option. The selection of topics has been based on those methods where technical information is freely available.

The ability to apply immunoassays to a multitude of widely differing analytes has been one of their major advantages. Ironically, it is also one of the complexities in producing a book like this one. Much of the specific nature of any particular analytical method will be determined by the individual physicochemical characteristics of the given analyte. For example, practical steps in sample preparation (e.g. concentration or purification), selection of appropriate calibrants, purification of a labelled analyte or antigen conjugates would all depend on the individual physico-chemical nature. Generally, this information should be sourced elsewhere.

1999 R.E.

Contents

Contributors

K. E. ANDERSON
The Babraham Institute, Babraham, Cambridge CB2 4AT

GEOFF BARNARD
Endocrine Unit, Department of Chemical Pathology, Southampton General Hospital, Tremona Road, Southampton SO16 6YD

DAVID BATES
Dako Diagnostics Ltd., Denmark House, Angel Drove, Ely, Cambridgeshire CB7 4ET

STUART BLINCKO
Netria, St. Bartholomew's Hospital, Royal Hospitals NHS Trust, London EC1A 7BE

DAVID G. BULLOCK
UK NEQASs for Clinical Chemistry and for Thyroid Hormones, Wolfson EQA Laboratory, Queen Elizabeth Medical Centre, PO Box 3909, Birmingham B15 2UE

DAVID B. COOK
Department of Clinical Biochemistry, The Medical School, Framlington Place, Newcastle upon Tyne NE2 4HH

SHEN DECUN
Department of Isotopes, Institute of Atomic Energy, Beijing 102413, China

RAYMOND EDWARDS
Netria, St. Bartholomew's Hospital, Royal Hospitals NHS Trust, London EC1A 7BE

IAIN HOWES
Netria, St. Bartholomew's Hospital, Royal Hospitals NHS Trust, London EC1A 7BE

J. LITTLE
Netria, St. Bartholomew's Hospital, Royal Hospitals NHS Trust, London EC1A 7BE

FINLAY MACKENZIE
UK NEQASs for Clinical Chemistry and for Thyroid Hormones, Wolfson EQA Laboratory, Queen Elizabeth Medical Centre, PO Box 3909, Birmingham B15 2UE

DAVID J. NEWMAN

South West Thames Institute for Renal Research, St. Helier Hospital, Wrythe Lane, Carshalton, Surrey SM5 1AA

CHRISTOPHER P. PRICE

Department of Clinical Biochemistry, St Bartholomew's and The Royal London Hospital School of Medicine and Dentistry, Turner Street, London E1 2AD

SHEN RONGSEN

Institute of Radiation Medicine, 27 Taiping Road, Beijing 100850, China

COLIN H. SELF

Department of Clinical Biochemistry, The Medical School, Framlington Place, Newcastle upon Tyne NE2 4HH

HANSA THAKKAR

Clinical Pathology, SmithKline Beecham Pharmaceuticals, The Frythe, Welwyn, Hertfordshire AL6 9AR

Abbreviations

Ab	Antibody
Ag	antigen
ALTM	all laboratory trimmed mean
ANS	Anilino-sulphonic acid
AP	alkaline phosphatase
B	bound fraction
BSA	bovine serum albumin
CDI	carbodiimide
CNBr	cyanogen bromide
CV	coefficient of variation
DEAE	diethylaminoethyl
DELFIA	dissociation-enhanced luminescence fluorimmunoassay
DMF	dimethyl formamide
DMSO	dimethylsulphoxide
DNA	deoxyribose nucleic acid
EDC	1-ethyl-3-(3-dimethylaminopropyl)carbodiimide
EDTA	ethylenediaminetetraacetic acid
EIA	enzyme-labeled immunoassay
ELISA	enzyme-labeled immunosorbent assay
EQA	external quality assessment
EQAS	external quality assessment schemes
F	free fraction
FIA	fluoroimmunoassay
FOD	final optical density
GMP	good manufacturing practice
HRP	horseradish peroxidase
HSA	human serum albumin
IA	immunoassay
IEMA	immunoenzymometric assay
IFMA	immunoflurometric assay
IMA	immunometric assay
IQS	internal quality control
IRMA	immunoradiometric assay
KLH	keyhole limpet haemocyanin
LIMS	laboratory information management system
MES	2-(N-morpholino)ethane sulphonic acid
NAD	nicotine adenine dinucleotide
NADH	reduced form of NAD
NADP	nicotine adenine dinucleotide phosphate
NHS	N-hydroxysuccinimide

NRS	normal rabbit serum
NSB	non-specific binding
NSS	normal sheep serum
OD	optical density
PBS	phosphate buffered saline
PEG	polyethylene glycol 6000
PETIA	particle enhanced turbidimetric immunoassay
PETINIA	particle enhanced turbimetric inhibition immunoassay
PFIA	polarization fluoroimmunoassay
pNPP	p-nitrophenylphosphate
PS	polystyrene
PVN	polyvinylnaphthalene
QA	quality assessment
QC	quality control
RIA	radiolabeled immunoassay
RNA	ribose nucleic acid
SD	standard deviation
TMB	tetramethylbenzidene
TOC	total oxidisable carbon

1

Principles of immunodiagnostic tests and their development; with specific use of radioisotopes as tracers

RAYMOND EDWARDS, STUART BLINCKO and IAIN HOWES

1. Introduction

Immunodiagnostic tests, or immunoassays used for diagnostics, have been used extensively in many scientific disciplines and in many different ways. Such tests encompass any analytical method which use antibodies as reagents, the results from which assist a diagnostic interpretation. The most obvious use has been found in clinical applications, but, the term can be applied equally to other areas of investigation, e.g. food spoilage agents, environmental pollution and illicit drug use. Thus in a wider sense, it would be applied to the assessment of any intrinsic factor associated with a state or condition to be ascertained.

The format of these tests has been equally varied; covering simple manual methods monitored by radioisotopes or enzymes; fully automated systems with integrated sophisticated detection; immunosensors; and 'dip-stick' tests even for home use as exemplified by the home pregnancy test. The wide repertoire of immunoassay types is undoubtedly a reflection of considerable interest shown by scientists from many different disciplines. A brief review of the historical development of immunoassays illustrates the multi-disciplinary contributions, beginning with the work of Krause on the reaction of soluble antigens and antiserum in the 1890s (1). In the early 1900s Bechold tried to discriminate between individual reactions of the several antigens and antisera in complex mixtures by exploiting diffusion in gels (2). In 1917 Landsteiner published work which was to have a most significant application in the subsequent development of immunoassays (3). Essentially he described the organic chemistry techniques which enabled the production of antibodies to non-immunogenic molecules. Small molecules could be coupled to weakly immunogenic proteins, such as albumin, which when injected into animals, such as rabbits, would lead to the production of antibodies with binding attributes showing specificity for the non-immunogenic part of the conjugate. This 'artificial conjugated antigen' was an important demonstration of the key 'hapten' principle later to prove invaluable in raising antibodies to an extensive range of small molecules used in many immunodiagnostic tests. The first publication describing the quantitative potential of an antibody reaction appeared in 1929 (4). Heidelberger and Kendall found that the mass of precipitated complex increased and eventually decreased in proportion to increasing concentrations of antigen when reacted with a fixed concentration of antibody. The 'precipitin curve' obtained with known amounts of antigen could be used to measure unknown quantities in experimental samples (by interpolation). In 1941 Coons, a histologist, demonstrated improved sensitivity in immuno-histochemical detection by coupling fluorophores to antibody molecules (5). This was to set a precedent for the many examples seen today for the use of fluorescent and chemiluminescent labels.

The technique of immunological analysis by gel diffusion was revisited in 1946 by Oudin (6) and 1947 by Ouchterlony (7). Their publications, which now represent the standard references in methods, describe simple and double diffusion in tubes or plates, respectively. Their work is still the basis for routine identification and quantification of large proteins such as specific immunoglobulin concentrations in serum.

Seemingly the first published reference to what might now be considered a conventional immunoassay was in 1954 by Statvitsky and Arquila (8). Their careful work clearly illustrated all the features of today's immunoassay, although perhaps full potential was constrained by their use of red blood cells as labels and haemagglutination for detection.

In the late 1950s, work in two centres, Ekins in London (9) and Berson and Yalow in New York (10), led to what must be considered a milestone in the advent of immunodiagnostic tests. Radioisotopes were used by both groups to label assay reagents and develop the first radioimmunoassays. These were to become routine analytical procedures for many substances in clinical diagnosis for the next two to three decades. The real significance of their work is seen in the apparent increase in analytical sensitivity (several orders of magnitude) over existing conventional methods. In 1967 Wide (11) and Miles and Hales (12) further improved sensitivity by introducing radiolabelled antibodies and the immunoradiometric assay. This was followed in the next year by two-site immunoradiometric assay (13, 14, and 15), using two separate antibodies. The need for purified antibodies with monospecific binding sites was fulfilled by the advances in monoclonal antibody production following the publication of Kohler and Milstein in 1975 (16). In recent years, the main developments are in the proliferation in different types of molecules or molecular attributes for use as labels or detection systems. The aim has been to improve sensitivity for simpler and earlier diagnosis or to minimize the amount of sample to be analysed.

Commercial competition has also accompanied the proliferation in immunoassay type leading to considerable constraints on technical information. Many examples are limited by either patented technology or undisclosed procedures. When these become integrated into 'black box' technology for fully automated analysis, they allow no access or interaction for general laboratory use. The techniques selected and described in this volume represent those that are accessible to the public domain and practically suitable for general laboratory development and application.

2. Classification of type

In principle all immunoassays reflect the basic reaction between antibody and antigen (analyte) as follows:

$$\underset{\text{'free'}}{\text{Ab} + \text{Ag}} = \underset{\text{'bound'}}{\text{Ab}\sim\text{Ag}}$$

Generally a fixed concentration of antibody reacts with varying concentrations of antigen forming an antibody–antigen complex ('bound' antigen). The reaction is reversible and will proceed to an equilibrium. The proportions of this can be described by a constant, K. As the value of K increases so too does the proportion of 'bound' complex. As far as the use of antibodies in immunodiagnostic tests is concerned there is an optimal K value for the measurement of any given concentration of antigen. It is generally recognized that as the target analyte concentration decreases, so the optimal K value needs to increase. This relationship is described in more detail later.

It is perhaps surprising that this simple relationship underlies the many varied types of immunodiagnostic tests. Nonetheless most attempts at designating type or format often seem unnecessarily complex and abstruse particularly when reduced to acronyms. The purpose seems primarily to give a unique identity, rather than to illustrate principle and relationship. The effect has been described as acronymic chaos (17)!

There is no consistent view on classification or terminology with respect to immunoassay type. Divisions can be made in a number of different ways, and given the plethora of variations possible it is not surprising that no coherent or universal scheme has been adopted. In considering a simple and definitive basis for classification, it is useful to invoke the 'precipitin curve' referred to earlier (4). This curve illustrates the principle subdivisions of immunoassay reactions as shown in *Figure. 1*.

The particular characteristics of the curve arise as a consequence of using a fixed and constant concentration of antibody (i.e. the analytical reagent) in all tests. Thus the change in shape results from the relative concentration or proportion of antigen (i.e. the analyte). The three main divisions can be summarized as follows:

- antibody concentration in relative excess
- antibody and antigen in relatively equal proportions
- antigen concentration in relative excess

In the first case using a fixed concentration of antibody in excess favours the total binding of antigen. Practically, a minute number of antibody binding sites will remain unoccupied. Here the change in response is directly proportional to the change in concentration of antigen. All formats conforming to this principle can be grouped together under the general heading of immunometric assays.

Secondly, where the antibody and antigen are present in approximately equal proportions, conditions favour the maximum formation of precipitates. This is because of inherent bivalency in the antibody molecule, i.e. two similar binding sites, and application to sufficiently large antigen molecules each providing at least two binding epitopes. Most of these techniques depend on the production of a precipitate seen as a visible line in a relatively transparent gel. In many examples the precipitate occurs following diffusion of one or both components until their concentrations reach equivalence (18). Electrophoresis can be used to assist diffusion, as in electroimmunoassay or the Laurell Rocket Technique (19). Whilst these techniques are often semi-quantitative they can be used to exploit differences in molecular charge or size and thus discriminate in a more qualitative way.

In the third division, antibody is present in a limited concentration with the antigen concentration in relative excess. As is clear in the illustrated precipitin curve the signal changes inversely to the concentration of antigen. Thus as the amount of antigen increases the degree of precipitation decreases. This is

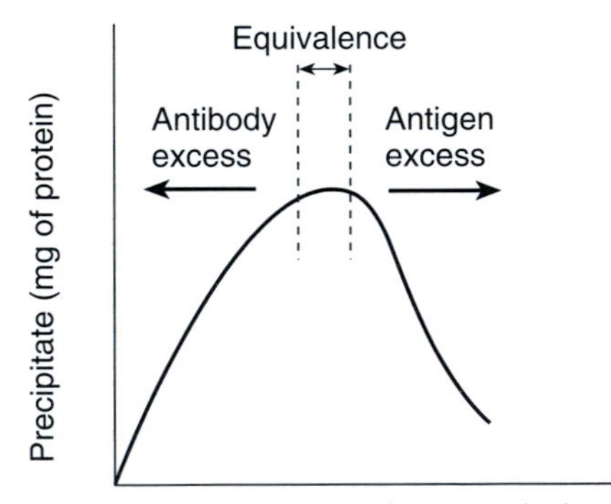

Figure 1. Precipitin curve; this traditional curve illustrates the relationship between a fixed amount of antibody and increasing amounts of antigen (analyte). Dividing the curve into three segments provides a basis for classifying immunoassays on the basis of relative concentrations.

directly analogous to the relationship found in immunoassays (IA) or competitive assays.

Overlaying these three basic divisions is the choice, where appropriate, of label component and subsequent detection system. The range of label components is illustrated in the following list:

- radioisotopes—especially ^{125}I, ^{3}H and ^{57}Co
- enzymes—notably horseradish peroxide or alkaline phosphatase
- fluorophores—simple, polarization and time-resolved modes
- chemiluminescence—e.g. luminol derivatives and acridiium esters
- particle/agglutination—exemplified by turbidimetry, nephelometry and particle counting

Others include bacteriophage/cytolysis; metal ions/atomic absorption; electron spin resonance etc.

Immunoassays have also been characterized on the basis of separation. A method for separating the antibody bound complex from unreacted components is usually an essential part of immunoassays. Many techniques have been used (20) ranging from crude non-specific methods based on protein precipitation, eletrophoretic charge or adsorption; to more sophisticated specific methods using second antibodies or solid-phase coupled antibodies. Although the type of separation can be a distinguishing feature, it is not really related to any fundamental principle and as such does not contribute much to a classification system.

There have been several examples of assays where the signal from the labelled component is moderated or significantly changed when binding between antibody and antigen takes place. Assays based on this principle and thus requiring no separation have been called homogeneous assays. By comparison any assay using separation of the bound and unreacted entities is called heterogeneous. Examples of homogeneous assays include modification of enzyme activity, reduction of polarized fluorescence, quenching of fluorescence and enhancement of fluorescence.

A rational system of classification is summarised in *Table 1*.

It is not within the scope of this publication to expound in detail the theory of immunoassay; however, this subject has been reviewed (21, 22 and 23). A glossary of terms frequently used in immunoassay literature is listed in *Table 2*.

3. Selection and preparation of reagents

There are many components in any diagnostic test. Some of these, such as sample preparation, selection of buffer constituents, additives and separation method, will be dealt with in Section 4—Assay design. In this section, the

Table 1.

Principle: Antibody(Ab) + Antigen(Ag) = Antibody complex–antigen(Ab:Ag)
[free] [bound]

Reaction follows law of mass action

Three types	**Immunometric assay (IMA)**		**Equivalence precipitation**	**Immunoassay (IA)**	
Proportion of reagent	Antibody in relative excess		Ab and Ag in relative equivalence	Antibody in limited amount (i.e. Ag in excess)	
Reagent label	Labelled antibody		No label (visual precipitate)	Labelled antigen	
Basic forms	IRMA—radioisotope IEMA—enzyme		Gel diffusion e.g. radial immunodiffusion (RID)	RIA—radioisotope EIA—enzyme	
	IFMA—fluorophore ICMA—chemiluminescence		Electrophoresis e.g. Laurell Rockets	FIA—fluorophore LIA—chemiluminescence	
	1 site IMA	**2 site IMA**		**Homogeneous**	**Heterogeneous**
		also called 'sandwich' use two distinct antibodies		non-separation usually EIA or FIA	separation

NB. Elisa is an acronym (enzyme labelled immunosorbent assay) usually applied to microtitre plate tests using enzyme labels—can be IEMA or EIA.

selection and preparation of the three key reagents will be outlined. These are:

- antibody or antiserum
- labelled tracer, either antigen or antibody
- calibrant or standard

3.1 Antibody or antiserum

Antibodies or immunoglobulins are a heterogeneous group of glycoproteins present in the serum and tissues and are produced by B-cell lymphocytes. Serum when it contains specified antibodies is referred to as antiserum.

The structure of all antibodies is based on a unit of four polypeptide chains: two identical 'light' chains and two identical 'heavy' chains, linked together by disulphide bonds in a distinctive conformation. Essentially they are bivalent with two identical binding sites.

Immunoglobulins are divided into classes, i.e. IgG, IgM, IgE, IgA and IgD, and subclasses, e.g. IgG_1, IgG_2, etc. The IgG class immunoglobulins in animal sera constitute about 75% of total serum immunoglobulins. Antibodies produced by hyperimmunization are predominantly IgG, with a small proportion of IgM. IgM antibodies are produced first during the primary antibody response. In practice they are often less stable.

3.1.1 Production Of antibodies

The initial stimulation of antibody production is usually achieved in an animal following the injection of an immunogen. Some antigens are immunogenic; that is, they will elicit an immune response, including the production of antibodies, when administered into an animal such as a rabbit. The key factors that confer immunogenicity are molecular size and defined tertiary structure. Generally, small molecules are less immunogenic than larger molecules, and it is likely that molecules smaller than about 3000 daltons are not immunogenic. If an antigen is either not immunogenic or only weakly immunogenic, it should be coupled to an immunogenic carrier, before administration (Section 3.1.4). Techniques for production of antibodies are:

- immunization for polyclonal antisera
- culture of monoclonal antibodies
- genetic engineering
- molecular imprinting

Although both genetically engineered antibodies (24) and molecular imprinting (25) have considerable potential to greatly extend the range and quality of antibodies, they have not yet been used practically in any diagnostic test.

Antisera may be raised successfully in several different animal species. Rabbit and sheep have been extensively used.

3.1.2 Production of polyclonal antisera

The administration of an immunogen *in vivo* will stimulate different cells of the immune system giving rise to a mixed population of antibodies derived from a number of B-lymphocyte clones (polyclonal). Serum taken from an immunised animal containing these antibodies is referred to as a polyclonal antiserum.

The immunisation of animals and subsequent handling is subject to regulations and legislation and it is best done by or under the guidance of an experienced expert. There are many commercial sources of polyclonal antisera and custom antiserum supply. A broad outline of an immunisation schedule is given in *Protocol 1*.

Table 2. Terms used in association with immunodiagnostic tests

Affinity	Strength of the association between antibody and antigen, a thermodynamic expression of primary binding energy
Allotype	Site shared in common by groups of immunoglobulins. Anti-allotypic antibodies are those that recognize and bind to common epitopes (in the constant region) of antibodies (see 'Analyte')
Analyte	That which is measured, often synonymous with the term antigen in immunoassays
Antigen	That which generated the antibody and specifically binds to the antibody binding site (see 'Analyte')
Avidity	Potential antigen binding capacity of antibody; related to affinity and number of binding sites
Biphasic response	See 'Hook'
Bound (fraction)	Proportion of analyte (or labelled analyte) bound to antibody (see 'Free')
Coating	Process of passive adsorption of antibodies
Complex	Usually refers to the antibody–antigen bound complex with multivalent binding
Conjugate	Often refers to an antigen or hapten covalently linked to a label, e.g. antigen–enzyme
Cross-reactivity	Relative potency (as %) of substances other than antigen (but usually sharing some molecular similarity) with the antibody
Double antibody	See 'Second antibody'
Epitope	That part of the antigen that reacts with the antibody binding site, i.e. antigenic determinant
Free (fraction)	Proportion of analyte (or labelled analyte) not bound to antibody (see 'Bound')
Free (hormone)	Refers to the measurement of that part of circulating hormone not bound by 'carrier' protein NB not to be confused with 'Free fraction'—see above.
Hapten	A small molecular entity which will bind with an antibody binding site but requires coupling to a larger molecule to elicit an immune response (i.e. to generate an antibody response)
Heterogeneous	Immunoassays requiring physical separation of bound and free fractions (see 'Homogeneous')
Heteroscedasticity	Non-uniformity of error in the response variable. This is a common feature of immunoassays
Homogeneous	Immunoassays where bound or free fractions are discriminated without separation (see 'Heterogeneous')
Hook	Dose response relationship that inverts (i.e. 'hooks'), sometimes referred to as 'biphasic'. This phenomenon is seen in some immunometric assays (IMAs) and could give rise to misleading results
Idiotype	Unique site on individual immunoglobulin molecules. Anti-idiotypic antibodies are those that recognize and bind a part (in the hypervariable region) of a unique antibody (e.g. bind to the hapten binding site (see 'Allotype'))
Label	A synonym for labelled antigen. The antigen or at least that molecular entity that binds to a specific binding site 'labelled' with a moiety to facilitate detection (e.g. radioisotope, enzyme, fluorophore, etc.)
Matrix	Medium used as a base for making standards (see 'Standards') or quality control pools

Table 2. (*Continued*)

Misclassification error	The measurement of part of the bound fraction as free or vice versa
Non-specific binding (NSB)	The proportion of 'labelled' antigen (see 'Label') that is measured in the bound fraction but is not bound to the specific binding site, such as tube surfaces, interstices of precipitate, etc.
Second antibody	An antibody raised in a 'second' species against the immunoglobulins of the primary species (e.g. donkey anti-rabbit). Usually used to form a precipitable complex
Standard	Usually refers to the material used to calibrate the assay, hence 'standard' curve when referring to dose-response relationship
Titre	The dilution of an antiserum giving a specific response (e.g. 50% binding of labelled antigen)
Tracer	A synonym for 'labelled' antigen (see 'Label')
Zero binding	Binding of tracer in the absence of analyte, often referred to as B_0

Protocol 1. General method for polyclonal antiserum in sheep[a]

Method

1. Dissolve 150 μg immunogen into 2 ml of normal saline.
2. Add to 4 ml of Freund's complete or incomplete adjuvant.[b]
3. Prepare a stable water in oil emulsion.
4. Inject 2 ml of emulsion into four sites subcutaneously per animal.
5. Bleed the animal approximately 3 weeks later for testing.
6. Inject booster doses every 4–8 weeks until satisfactory antiserum is obtained.

[a]Reduce volumes for smaller animals e.g. rabbit.
[b]The complete adjuvant containing killed *Mycobacterium butyricium* to be used only for the primary injection. The incomplete adjuvant is used for all subsequent injections. The adjuvants contain mixtures of paraffin oil (e.g. Boyol F) and detergent (e.g. mannide monooleate).

3.1.3 Production of monoclonal antibodies

Monoclonal antibody production has rapidly become established practice, although involving a variety of approaches. The original method of Kohler and Milstein (16) was to fuse B-cell antibody producers with neoplastic tumour cells (B-cell myelomas), producing immortal hybrid cells or 'hybridomas'. Neoplastic cells were chosen to have two properties:

1. The cells should be killed by a selective medium in which hybridomas survive. The usual medium contains hypoxanthine, aminopterin and thymidine (HAT), which selectively kills cells deficient in certain enzymes.
2. The tumour cells should not secrete an antibody, which would contaminate the final monoclonal antibody (mab), and preferably not even immunoglobulin light chains, as these may combine with the mab heavy chain producing inactive hybrid molecules.

Several HAT-sensitive, non-secreting B-cell lines have been isolated from BALB/C strain mice. More recently suitable sheep and human lines have been prepared. For reasons of history and convenience, most monoclonal work has been with the mouse. From a practical point of view, it is usually preferable to source mabs from the large number of commercial suppliers, or to consider custom production on a contract basis. The 'in-house' development of mab is an expensive undertaking requiring considerable investment of time, money and expertise with speculative results.

The conventional approach using immunized mice and mouse derived myeloma cell lines consists of four aspects:

(1) immunization;
(2) cell fusion;
(3) cloning;
(4) culturing and purification.

Immune cells are harvested from the animal following an immunization schedule and demonstration of satisfactory test-bleeds. In the case of mice, the animals are usually sacrificed and spleen cells prepared aseptically. The latter are induced to fuse with the chosen tumour cell by mixing at high density in the presence of polyethylene glycol (50%). The efficiency of fusion is generally low, e.g. 0.1%, but enough hybrid cells are produced which survive in HAT medium. Detailed protocols for monoclonal production are given elsewhere (26).

3.1.4 Conjugated haptens

When an antigen is not immunogenic or only weakly immunogenic, it is necessary to couple it to an immunogenic carrier in order to effect an antibody response. The main factor governing the potential immunogenicity is molecular size. The following rule of thumb can be used to assess immunogenicity, and hence the need to couple to a carrier.

(a) Mol. wt. < 3000—probably not immunogenic, couple to carrier.
(b) Mol. wt. 3000–5000—weakly immunogenic, coupling to carrier should enhance immunogenicity.
(c) Mol. wt. 5000–10000—probably immunogenic, coupling may enhance immunogenicity.
(d) Mol. wt > 10000—should be immunogenic, usually no need to couple to carrier.

Various immunogenic carriers have been used, the most common are:

- bovine serum albumin (BSA) or human serum albumin (HSA)
- rabbit thyroglobulin
- keyhole limpet haemocyanin (KLH)

KLH and thyroglobulin are considered to be better carriers for small molecules and to give rise to antibodies with high affinity. However, BSA has been found to be satisfactory for many peptides. Three general methods are given: glutaraldehyde to couple via amine groups (*Protocol 2*), the mixed anhydride reaction (*Protocol 3*) to specifically activate carboxyl groups before reacting with amine groups, and the carbodiimide reaction (*Protocol 4*) to couple carboxyl groups and amine groups. These coupling reactions can lead to cross-linking of either antigen or carrier.

The molar ratio of coupled peptide to carrier is important for the production of good antisera. Satisfactory molar ratios are at least 10:1 for BSA; 20:1 for thyroglobulin; 80:1 for KLH (i.e. hapten:carrier).

Conjugates can be characterized by difference UV spectra, particularly if the hapten has an absorbance spectrum different from the carrier protein.

Protocol 2. Conjugation of a hapten to a carrier using glutaraldehyde

Method

1. Dissolve 1–2 mg hapten in 500 ml 0.1 M $NaHCO_3$.
2. Add 2–3 mg thyroglobulin (Sigma) to the above solution.
3. Add 10 ml glutaraldehyde (Sigma) aq. 25% (v/v).
4. React in the dark for 30 min.
5. Add 100 ml 0.5 M glycine.
6. Dialyse against 0.1 M $NaHCO_3$.
7. Store dialysed conjugate at –20 °C or lyophilize.

Protocol 3. Conjugation of hapten to carrier using mixed anhydride method

Method

1. Dissolve 3 mg hapten in 30 ml fresh anhydrous dioxane[a] and cool to 8–10 °C.
2. Add 7.5 ml tri(*n*)butylamine,[a] diluted 1:10 in dioxane.
3. Cool to 4 °C and add 5 ml isobutylchloroformate,[a] diluted 1:10 in dioxane.
4. React for 20 min at 4 °C.
5. Add 4 mg BSA in 200 ml of H_2O/dioxane (1:1) containing 40 ml 1 N NaOH. (NB Add NaOH to BSA in H_2O before adding dioxane.)
6. Stir for 4 h at 4 °C.
7. Dialyse against distilled H_2O.
8. Store conjugate at –20 °C or lyophilize.

[a] All solvents (Sigma) are used when fresh and dry.

Protocol 4. Conjugation of hapten to carrier using carbodiimide method

Method

1. Dissolve 20 mg peptide in 500 ml 50% aqueous pyridine.
2. Add 20 mg BSA to the above solution.
3. Add 100 mg 1-ethyl-3-(3-dimethylaminopropyl)carbodiimide hydrochloride (Sigma) in 250 ml 50% aqueous pyridine.
4. React at ambient temperature for 5 to 30 min.
5. Dialyse against distilled water.
6. Store conjugate at –20 °C or lyophilize.

3.2 Selection of antibody or antiserum

Characteristics of the antibody determine the potential performance of any immunodiagnostic test. Therefore it is important to fully characterize antibodies or antisera before choosing an appropriate one for use as an immunoreagent and proceeding with the development of a test.

Antibodies are characterized by three factors, namely:

(1) affinity;
(2) concentration of binding sites;
(3) specificity.

The most practical tests to assess these factors are:

1. Titre—a dilution curve plotting the binding of a small fixed concentration of antigen. This gives a measure of both concentration and affinity.
2. Scatchard analysis of binding curve—a plot of binding against concentration of antigen which gives details of both affinity and number of binding sites.
3. Cross-reaction studies, usually with structurally related molecules to give information on the type and degree of specificity.

3.2.1 Determination of titre

The titre has been used extensively as a measure of quality. As a parameter it reflects both concentration and affinity of the antibodies. In practice, antisera with high titres often exhibit high binding affinities. Satisfactory affinity of binding can be demonstrated using the same antiserum dilution curve as for the titre repeated in the presence of a specific concentration of antigen. Displacement between the two curves denotes adequate sensitivity which in turn denotes an appropriate affinity (see *Figure 2*).

Protocol 5. Procedure for determining titre

Method

1. Dilute antiserum culture medium or ascitic fluid in assay buffer using serial dilutions of tenfold from 1:100 to 1:1000000 (or more).
2. Dilute labelled antigen in assay buffer to give minimum concentration compatible with precise measurement (e.g. for ^{125}I-tracer use 200000 cpm per ml).
3. Dilute antigen (unlabelled) in assay buffer to give the minimum concentration that must be detected (i.e. required sensitivity).
4. React the following:
   (i) 100 μl of diluted antiserum
   (ii) 100 μl of label
   (iii) 100 μl of either assay buffer i.e. 'zero' antigen or 'low' antigen solution, until equilibrium is reached e.g. overnight at room temperature.
5. Add separation reagent, e.g. solid-phase second antibody—100 μl of suspension (IDS Ltd), react for appropriate time with shaking or mixing to ensure all bound fraction is completed. Centrifuge at 1000 g for 10 min. Alternative separation—add 50 μl diluted second antiserum (+ carrier serum), react for 1 hour and add 1 ml 4% Polyethylene Glycol 6000 (PEG 6000) containing 0.1% Triton X-100.
6. Decant and wash solid-phase precipitate with 2 ml assay buffer, repeating centrifugation and decantation. (Do not wash AB_2/PEG precipitate.)
7. Measure bound activity e.g. count radioactivity. Calculate per cent bound (%B) for each antiserum dilution taking maximum binding as that achieved with excess antibody. NB Subtract non-specific binding.
8. Plot %B (*y* axis) against antiserum dilution (*x* axis).Titre is given by the antiserum dilution to give 50% of maximum, i.e. 50%B. Maximum displacement between the two curves gives optimal dilution of antiserum for sensitivity.

Assay buffer—50 mM phosphate buffer pH 7.4 containing 150 mM saline, 0.1% albumin and 0.1% sodium azide.

3.2.2 Scatchard analysis

Scatchard analysis of binding data is derived from a ratio of specifically bound antigen to free antigen (B/F) plotted against the concentration of specifically bound antigen ([B]). This plot is capable of giving an estimate of the binding

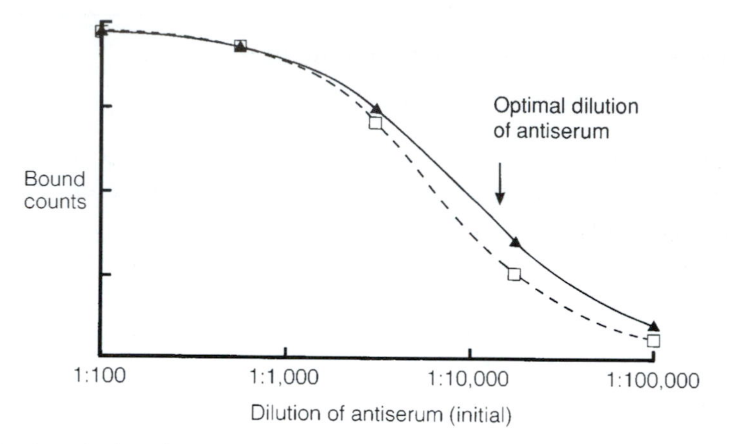

Figure 2. A typical antibody dilution curve to calculate titre; the percentage of bound tracer (in terms of radioactive counts) is plotted against solutions of antiserum. Titre is given by the dilution that would give 50% bound (of Bo). A second curve obtained in the presence of the minimal quantity of analyte indicates adequate sensitivity.

affinity (i.e. K or equilibrium constant) and the number of available binding sites per unit volume.

Protocol 6. Procedure for deriving a Scatchard plot

Method

1. Dilute antigen in assay buffer to cover a range of concentrations (approximately 500-fold) including the lowest concentration for sensitivity.
2. Dilute antiserum or antibody solution in assay buffer to give approximate dilution indicated from titre studies, i.e. 50% maximum binding or maximum displacement of titre curves.
3. Dilute labelled antigen in assay buffer to give minimum concentration to achieve detection with reasonable precision (e.g. for ^{125}I-tracer use approximately 200000 cpm per ml). NB It will be necessary to ensure that the precise concentration of labelled antigen is known.
4. React all reagents (e.g. 100 μl of various concentrations of antigen, 100 μl of labelled antigen and 100 μl of antibody) until at equilibrium (e.g. overnight at room temperature).
5. Add separation agent, e.g. solid-phase second antibody—100 μl of suspension of Sac-cel (IDS Ltd)—react for appropriate time with shaking or mixing to ensure all bound fraction is completed, e.g. 30 min. Centrifuge at 1000 g for 10 min. Alternative separation—add 50 μl of diluted second antiserum, react for one hour and add 1 ml 4% PEG (6000). Centrifuge at 2000 g for 30 min.

6. Decant and wash solid-phase precipitate with 2 ml assay buffer containing 0.1% TritonX-100 (not necessary to wash AB_2/PEG precipitate).
7. Measure bound activity (e.g. count radioactivity). Calculate the following for each concentration:
   - B—specifically bound labelled antigen after subtracting non-specific binding
   - F—calculated by subtracting B from total activity determined in presence of excess antibody, dividing B by F for each concentration
   - [B]—calculate concentration of specifically bound antigen from total for antigen + total for labelled antigen x %B (B as percentage of total), for each concentration. Plot B/F as *y* variable and [B] as *x* variable.
8. Fit straight line through data points. Calculate gradient to find—*K* (equilibrium constant for the antibody—antigen reaction). The intercept on the *x* axis gives the total concentration of binding sites measurable under assay conditions. In a mixed population of antibodies Scatchard analysis may yield a curve which needs resolution into two or more straight lines (see *Figure 3*).

Assay buffer—50 mM phosphate buffer pH 7.4 containing 150 mM saline, 0.1% albumin and 0.1% sodium azide.

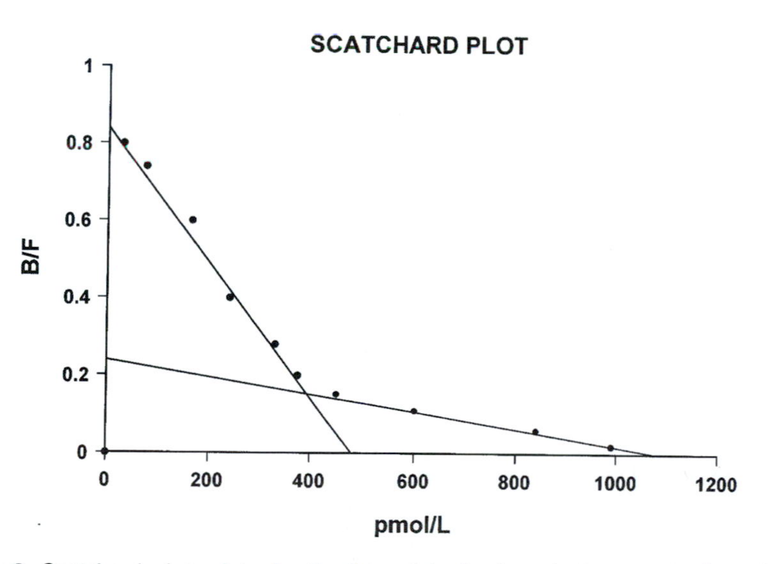

Figure 3. Scatchard plot; plot of ratio (bound to free) against concentration of found analyte. Gradient of the slope gives *K* and intercept (*x* axis) gives number of binding sites. This example indicates the presence of two antibody populations, high *K* value = 2×10^9 litre mol^{-1} and lower *K* value = 2.4×10^8 litre mol^{-1}.

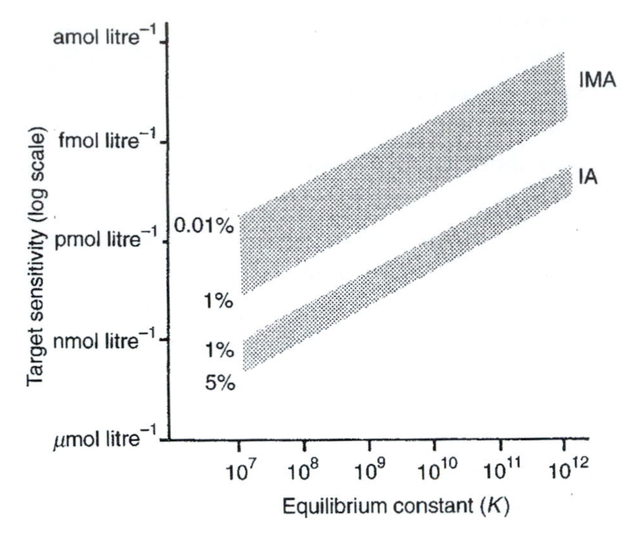

Figure 4. Nomogram simplifying relationship between target sensitivity and appropriate K value. Details are given for both immunoassays (i.e. competitive, IA) and immunometric assays (i.e. excess reagents, IMA). The proportion of signal due to non-specific binding (NSB) affects the relationship as shown; percentage figures indicate NSB.

The most important characteristic in selecting an antibody is this equilibrium (affinity) constant K. Antibodies with higher K value will give better sensitivity. There is a direct relationship between K value and potential sensitivity, which is modified by both assay format (see Section 4) and the NSB value. These relationships are summarized in diagrammatic form in *Figure 4*.

3.2.3 Cross-reactions

Molecules structurally related to each antigen are used to determine the potential specificity of any given antiserum or antibody. Each cross-reactant will give a relative potency in terms of binding compared with the specific antigen. Appropriate selection of cross-reactants differing in various parts of the molecule indicates where on the molecule the antibody is most specific for.

Protocol 7. Procedure for determining cross-reaction

Method

1. Prepare the reagents as for Scatchard analysis.
2. Also prepare various dilutions of cross-reactants in assay buffer, usually at higher concentrations than for antigen.

3. React the following;
   (i) 100 μl of antigen at various concentrations or 100 μl of cross-reactant at various concentrations
   (ii) 100 μl of labelled antigen
   (iii) 100 μl diluted antiserum or antibody until equilibrium is reached, e.g. overnight at room temperature.
4. Separate bound fractions and measure bound activity as in steps 5, 6 and 7 of Scatchard analysis (*Protocol 6*).
5. Plot bound activity as percentage of maximum binding at 'zero' concentration (%Bo) on *x* axis against concentration on *y* axis. Cross-reaction is given by the relative potency in terms of concentration to give 50% Bo compared with the concentration of antigen to give 50% Bo. For example, if 10 nmol/litre of cross-reactant and 1 nmol/litre of antigen both give 50% Bo, then the relative potency for cross-reactant is 10%.

3.3 Purification of antibodies and antisera

Antisera have been used most commonly in an unpurified form. The dilution factor is frequently high enough so that other components in the serum are practically ineffective. However, some degree of purification will sometimes confer benefits, particularly for preparation of solid-phase reagents, and in some cases is necessary for satisfactory performance, as in the preparation of labelled components. The development of monoclonal antibody technology has greatly simplified the purification of specific immunoreagents.

The choice of method for purifying an antibody preparation will depend on the degree of priority required, the matrix containing the antibody, e.g. serum or culture medium, and the type of immunoglobulin.

The most common methods that cover all necessary situations are summarized as follows:

- immunoglobulin (IgG) enriched preparations using salt fractionation
- IgG fraction using caprylic (octanoic) acid
- ion-exchange chromatography
- affinity chromatography specific for Fc potion using protein A and G
- affinity chromatography using specific immunoabsorbants
- preparation of purified antibody fragments; Fab and $F(ab)_2$

3.3.1 IgG-enriched preparations from serum

Salt fractionation

Conventionally this is achieved with either ammonium or sodium sulphate. The method makes use of the fact that immunoglobulins precipitate at lower ionic strength than most other serum proteins. Although the product obtained contains unwanted proteins, many others such as albumin remain in solution. This 'salting out' is useful as a prelude to further purification and also for concentration of dilute antibody solutions.

Protocol 8. Ammonium sulphate (40–45% saturation) precipitation of IgG

Method

1. Place antiserum in a beaker and stir with a magnetic stirrer. The temperature may be between 4 and 25 °C[a]; for more labile antibodies a lower temperature is recommended.
2. Add 2.4–2.7 g[b] of solid ammonium sulphate, $(NH_4)_2SO_4$ (e.g. Ammonium Sulphate for Biochemistry, from Merck), for every 10 ml of serum. The solid must be added slowly, e.g. 4 g per minute, to avoid high local salt concentrations.
3. Incubate with stirring for 30 min.
4. Centrifuge at $>$ 3000 g for 30 min at constant temperature.
5. Discard the supernatant and wash the pellet with 40% saturated ammonium sulphate (240 g/litre in distilled water). This reduces contamination by non-immunoglobulin protein. Careful mixing is necessary as the solution is viscous.
6. Recentrifuge the suspension (step 4) and repeat the wash and centrifugation once more.
7. Dissolve the precipitate in distilled water; approximately 2 ml per 10 ml original serum.
8. For storage, dialyse the protein solution against phosphate buffered saline (PBS, 25 mM sodium phosphate, pH 7.4/150 mM NaCl) containing 0.1% sodium azide (w/v) (four changes of 1000 volumes). Keep at 4 °C. For long-term storage freeze in convenient aliquots at –70 °C.
9. Dialyse against an appropriate buffer where further purification is to be carried out.

[a] Solubility of $(NH_4)_2SO_4$ changes only slightly between 0 °C and 25 °C.
[b] Most IgG precipitates at 40% saturation (2.4 g/10 ml) but it may be necessary to go to 45% to collect all antibody. Contamination with other proteins is greater at 45%.

Whilst not as mild as the ammonium sulphate method, sodium sulphate precipitation gives a purer preparation, commonly referred to as an 'IgG cut'.

Protocol 9. Sodium sulphate precipitation of IgG

Method

1. Warm antiserum to 25°C[a] and stir in a beaker by magnetic stirring or mix on a vortex mixer.
2. Add 18 g anhydrous Na_2SO_4 (e.g. AnalaR, Merck) per 10 ml serum slowly to give 18% (w/v) solution. Continue to mix during addition.
3. Incubate for 30 min at 25°C with stirring.
4. Centrifuge at $>$ 3000 g for 30 min at 25°C (temperature controlled centrifuge).
5. Collect the precipitate and note its volume. The pellet should be redissolved in distilled water to give 5 ml per 10 ml original antiserum.
6. Warm the solution to 25°C and gradually add Na_2SO_4 to give 14% (w/v) salt, allowing for that retained in the precipitate.
7. Recentrifuge as in step 4. Decant the supernatant solution and redissolve the precipitate in about 3 ml distilled water per 10 ml original antiserum.
8. For storage, dialyse the protein solution against phosphate buffered saline (PBS, 25 mM sodium phosphate, pH 7.4/150 mM NaCl) containing 0.1% sodium azide (four changes of 1000 volumes). Keep at 4°C. For long-term storage freeze in convenient aliquots at–70°C.
9. Dialyse against an appropriate buffer where further purification is to be carried out.

[a]Use a thermostatted water bath. The solubility of Na_2SO_4 is very temperature dependent.

Octanoic acid (caprylic) precipitation of antiserum

In this procedure the bulk of serum protein is made to precipitate leaving IgG in solution. This has the advantage of avoiding damage to the antibody due to salt precipitation followed by redissolution. The method uses octanoic (caprylic) acid in mildly acidic conditions (pH 4.8), near to the p*I* of the majority of serum proteins. Hydrophobic interactions between uncharged proteins are increased by the aliphatic carboxylic acid, producing a precipitate.

Protocol 10. Octanoic (caprylic) acid precipitation of antiserum

Method

1. To one volume antiserum add two volumes sodium acetate buffer, 60 mM, pH 4.0[a], and mix at room temperature (21 °C) on a magnetic stirrer.[b] pH should be about 4.8.
2. To the stirred mixture add 0.68 g *n*-octanoic acid (e.g. Merck) per 10 ml original antiserum dropwise.
3. When all octanoic acid has been added continue to stir for 30 min.
4. Centrifuge at > 3000 g for 30 min at 21 °C.
5. Collect the supernatant containing the IgG. If a lipoprotein layer is present at the surface of the solution avoid or remove it by filtration with glass fibre paper.
6. If desired, antibody yield can be increased by elution of the pellet with acetate buffer (2 volumes) adjusted to pH 4.8.[c] Collect by centrifugation (step 4).
7. Pool supernatant solutions. For storage, dialyse against phosphate buffered saline (PBS, 25 mM sodium phosphate, pH 7.4/150 mM NaCl) containing 0.1% sodium azide (four changes of 1000 volumes). Keep at 4 °C. For long-term storage freeze in convenient aliquots at –70 °C.
8. Dialyse against an appropriate buffer where further purification is to be carried out.

[a]Dissolve 3.6 g glacial acetic acid per litre distilled water and adjust the pH to 4.0 with 1 M NaOH.
[b]Use clean glassware throughout.
[c]With 1 M NaOH.

3.3.2 Ion-exchange purification of IgG

Following salt or octanoic acid precipitation, the IgG fraction may be isolated using diethylaminoethyl (DEAE) ion-exchange chromatography. At pH 8.0, DEAE groups are positively charged and are balanced by negatively charged anions. Virtually all proteins are negatively charged at pH 8.0 and bind to immobilized DEAE groups. However, immunoglobulins are relatively basic (having p*I* between 7.0 and 8.0) and, as either the pH is lowered or the concentration of competing anions (e.g. Cl^-) increases, are the first proteins to be released from the ion-exchange matrix.

Conventionally, DEAE–cellulose has been widely used as the ion-exchange medium. The method given in *Protocol 11* uses DEAE–Sephacel (beaded cellulose) which need not be 'recycled' before use and which has superior flow characteristics. The method described may be used for both polyclonal and monoclonal.

Protocol 11. Ion-exchange chromatography

Method

1. Equilibrate approximately 100 ml of DEAE–Sephacel (Pharmacia) by washing four times with 1l 10 mM Tris–HCl buffer, pH 8.0, on a glass sinter.
2. Prepare a column of suitable size—not less than 10 ml slurry is required for 100 mg protein.[a] Approximate dimensions should be length = 20× diameter, e.g. 1.6 cm × 30 cm. Pour under gravity.
3. Load the column with material previously exchanged into 10 mM Tris–HCl by dialysis[b] at 50 ml/h.
4. Wash the column with three column volumes of Tris buffer.
5. Elute the column. This may be done by stepwise or linear NaCl concentration gradient:
   (i) Stepwise gradient[c]—running at 50 ml/h elute with 50, 100, 150 and 200 mM NaCl, all prepared in 10 mM Tris–HCl, pH 8.0. Each step should be between 2 and 3 times the column volume.
   (ii) Linear gradient[d]—prepare a linear 0–300 mM NaCl gradient: use five column volumes each of 0 NaCl and 300 mM NaCl with a suitable gradient making apparatus.

[a] Protein content of IgG cuts etc. may be estimated by absorbance assuming $E_{280}^{1\%} = 14$.
[b] Gel filtration is not recommended because some immunoglobulins will precipitate at low ionic strength. Precipitated material may still be loaded onto the DEAE–Sephacel.
[c] Most IgG is desorbed by 200 mM NaCl; a higher concentration is only rarely needed. Overlap of immunoglobulin from step to step often occurs. A long 'trailing edge' is common.
[d] Better yields and resolution are obtained with a continuous gradient.

3.3.3 Protein A and Protein G chromatography of IgG

Protein A from *Staphylococcus aureus* (MW = 42000) has a strong affinity for the Fc portion of the IgG molecule. One molecule of Protein A can bind two IgG molecules. However, the strength of binding is dependent on the species and subclass—see *Table 3*—and some IgGs do not bind at all.

Protein G from Group G streptococci is a type III Fc receptor which binds IgG in a similar fashion to Protein A. Native Protein G bears an albumin receptor; this is deleted in the commercially produced recombinant ligand (MW = 17000). It has the advantage of binding IgGs unreactive with Protein A and generally has higher affinity for IgG.

Protein A and G may be used to isolate polyclonal IgG from antisera and mab from ascites fluid, serum or from dilute culture supernatants. Protein G is the method of choice for isolation of mab and IgG from rat, sheep and horse. Neither bind avian IgG. Because of the spread of affinity of Protein A for

Table 3. Relative binding affinities of Protein A and Protein G for IgG

Species	Protein A	Protein G
Mouse	+	+
Sheep	–	++
Horse	–	++
Rabbit	++	++
Human	++[a]	++

++ high affinity; + medium affinity; – insignificant affinity.
[a] IgG_3 is not bound.

different IgG subclasses, they can be partly resolved using Protein A chromatography.

Protein A and Protein G are available in a variety of ready to use affinity media and as prepacked columns.

Protocol 12. Protein A isolation of IgG

Method

1. Pack 2 ml Protein A Sepharose 4 Fast Flow (Pharmacia product 17–0974–01) containing approximately 24 mg of Protein A, into a column.
2. Wash the column with sodium citrate buffer (0.1 M, pH 3.0, 5 ml) followed by sodium phosphate (0.1 M, pH 8.0, 10 ml).
3. Bring the pH of 2 ml of antiserum or ascites fluid[a] to 8.0 with Tris base (2 M) and add 2 ml of phosphate buffer, mix and centrifuge (3000 g for 30 min).
4. Load the sample at about 5 ml/h, collecting 1 ml fractions.
5. Wash with phosphate buffer. The effluent contains non-immunoglobulin protein and non-adherent immunoglobulins.
6. Elute the column with 10 ml 0.1 M sodium citrate buffer.[b,c] After elution wash the column with 10 ml phosphate buffer.
7. Monitor protein elution by absorbance at 280 nm.
8. Dialyse the eluted fraction against phosphate buffered saline (PBS, 25 mM sodium phosphate, pH 7.4/150 mM NaCl) containing 0.1% sodium azide (four changes of 1000 volumes). Keep at 4°C. For long-term storage freeze in convenient aliquots at –70°C.

[a]For extraction of mab from culture supernatant use Protein G. Partially purified preparations of IgG may be substituted (see Section 4.1 and 4.2).
[b]IgG_1 is eluted at pH 6.0, IgG_{2a} at 4.5, IgG_{2b} at 3.5. All antibody can be eluted in a single step with pH 3.0.
[c]For pH 4.5 and below add 50 ml Tris base (2 M) to the fraction collection tubes.

Protocol 13. Protein G isolation of mab or polyclonal IgG

Method

1. Make a column of 1 ml Protein G Sepharose 4 Fast Flow (Pharmacia). This contains 2 mg Protein G and has a typical binding capacity of 20 mg IgG.
2. Wash the column with 5 ml 0.02 M sodium phosphate, pH 7.0, at a flow rate of 0.5 ml/min.
3. Load the antibody sample.[a] Adsorption of antibody is volume independent—flow rate 0.5 ml/min; collect 0.5 ml fractions.
4. Continue washing with 10 ml phosphate buffer.
5. Elute bound protein with 10 ml 0.1 M glycine–HCl buffer, pH 2.7, monitoring protein by absorbance at 280 nm. Neutralize antibody-containing fractions with 2 M Tris base (25 ml/0.5 ml).
6. Return the column to pH 7.0 by washing with a further 10 ml phosphate buffer.
7. Dialyse the eluted antibody against phosphate buffered saline (PBS, 25 mM sodium phosphate, pH 7.4/150 mM NaCl) containing 0.1% sodium azide (four changes of 1000 volumes). Keep at 4°C. For long-term storage freeze in convenient aliquots at –70°C.

[a]Antiserum, ascites fluid, culture supernatant or a partially purified preparation from any of these at pH 7.0. Remove particulate material by centrifugation.

Protocol 14. Immunoaffinity chromatography of polyclonal antisera

Equipment and reagents

- Cynanogen bromide (CNBr) activated Sepharose 4B (Pharmacia)
- Sintered glass filter
- Peristaltic pump with appropriate tubing for 6–20 ml/h
- Fraction collector
- Spectrophotometer (for 280 nm)
- All operations at ambient temperature
- Glass chromatography column (1 × 10 cm)

Method

A. Preparation of immunosorbent

1. Add CNBr activated Sepharose 4B in 1 mM HCl (approximately 1 g in 10 ml) to make a slurry and wash on sintered glass filter with a further 190 ml over 15 minutes.

Protocol 14. *Continued*

2. Dissolve antigen or hapten–albumin conjugate (1–2 mg per g of immunosorbent in coupling buffer:100 mM ammonium bicarbonate and 500 mM sodium chloride, pH 8.0). NB. No azide in buffer.
3. Add antigen solution to swollen gel and mix gently (by rotation) for 2 h.
4. Separate gel on glass filter.
5. Add 1 M ethanolamine pH 8 (Merck) in coupling buffer and mix gently for a further 2 h.
6. Repeat separation step to remove ethanolamine solution.
7. Wash prepared immunosorbent with the following cycle of buffers
   (a) 0.1 M sodium acetate trihydrate (Merck), pH 4.0 containing 1 M sodium chloride. NB. No azide.
   (b) 0.1 M di-sodium tetraborate (Merck), pH 8.0 containing 1 M sodium chloride. NB. No azide. Use approximately 30 ml per g gel.
8. Repeat this cycle twice more.
9. Wash in 50 mM phosphate buffer, pH 7.4 and store at 4°C in same phosphate buffer.

B. Affinity chromatography

1. Absorb specific antibodies onto gel by mixing with antiserum, volume depending on titre, approximately 10 ml per g. Mix over 1–2 days.
2. Transfer immunosorbent to column and wash with phosphate buffered saline 50 mM pH 7.0, 0.15 M sodium chloride, at approximately 20 ml per h until OD at 280 nm has returned to baseline.
3. Elute with stepwise gradient (approximately 15 ml per step) as follows:
   (a) 0.5 M sodium acetate pH 7.0 containing 20% acetonitrile.
   (b) 0.05 M sodium acetate pH 5.0 (adjust pH with acetic acid) containing 20% acetonitrile.
   (c) 0.05 M sodium acetate pH 4.0 (adjust pH with acetic acid) containing 20% acetonitrile.

   Collecting 1 ml factions into tubes containing 0.5 M phosphate, pH 7.4 (approximately 0.1 ml).
4. Monitor OD 280 nm and immunoreactivity to identify antibodies. Pool fractions (if appropriate) and dialyse to remove acetonitrile against neutral buffer, e.g. 0.05 M phosphate, pH 7.4.

3.3.4 Preparation of antibody fragments; Fab, $F(ab)_2$ and FC fragments

The preparation of antibody fragments is a useful option in that immunoreactivity, e.g. the binding site in the Fab portion, is retained whilst removing the potential for interference, e.g. interference from the Fc portion. The $F(ab)_2$ is divalent as it retains both binding sites in one fragment. Conventionally the enzyme pepsin is used to digest immunoglobulin molecules to give $F(ab)_2$ and papain for Fab and Fc fragments.

It is important to note that different classes of immunoglobulins differ in their sensitivity or resistance to enzyme digestion. This is also true for classes or subclasses from different species of animals. For example, sheep immunoglobulins are more resistant to pepsin than those from rabbit. Subclasses of mouse immunoglobulins differ widely in sensitivity to pepsin, in particular with respect to myeloma antibodies, viz. monoclonal antibodies. For example, in the mouse, IgG_3 is more sensitive than IgG_{2a} which is more than IgG_1. IgG_{2b} is very difficult.

Although the optimal conditions for enzyme digestion for each particular antibody or immunoglobulin batch must be arrived at empirically, the following protocols (15–17) have been given for guidance.

Protocol 15. General method to prepare $F(ab)_2$ fragments using pepsin

Method

1. Dialyse purified immunoglobulin preparation against 0.1 M acetate buffer, pH 4.3.
2. Add crystalline pepsin (Sigma 1:60000) in ratio 1:50 Ig(w/w), check pH and adjust if necessary. Gently mix to dissolve pepsin.
3. React for 8–14 hours at 37 °C, then cool in ice bath.
4. Centrifuge to remove any precipitate.
5. Adjust pH to 8.0 with 1 M NaOH to inactivate pepsin.
6. Dialyse extensively against 0.01 M phosphate buffer pH 7.2 containing 0.15 M NaCl (PBS).
7. Purify on Sephadex G-150 (Pharmacia) eluting with PBS.

Protocol 16. General method for Fab fragment using papain

Equipment and reagents

- Water bath with shaker
- Centrifuge
- Papain Merck 16000
- L-Cysteine hydrochloride (Sigma)
- Iodoacetamide (Sigma)
- Ethylenediamine tetraacetic acid disodium salt (EDTA, Sigma)
- IgG either at IgG cut or purified by other means at a concentration of about 30 mg/ml in saline (0.9% NaCl)

Protocol 16. *Continued*

Method

1. Adjust pH of IgG solution to 7.0.
2. Transfer solution to an appropriate volume screw top glass vial.
3. Bring to a temperature of 37 °C by shaking in a water bath for 30 minutes.
4. Add EDTA to a final concentration of 0.8 mg/ml. Mix until dissolved.
5. Add papain corresponding to 16% of the IgG (w/w). Gently mix until dissolved.
6. Add L-cysteine hydrochloride dissolved in a minimum volume of saline (0.9% NaCl) to a concentration of 1.6 mg/ml.
7. Shake the solution at 37 °C for one hour and then allow to stand at 37 °C for a further 23 hours.
8. Add iodoacetamide to a concentration of 2.8 mg/ml and shake for 15 minutes at 37 °C.
9. Centrifuge (4500 rpm for 45 min) to isolate the precipitate (Fc). Decant the supernatant which contains Fab.
10. Dialyse against saline (0.9% NaCl).
11. For further purification affinity chromatography (with immobilized specific antigen) or ion-exchange may be employed. For IgG from species other than sheep the Fc may be removed by Protein A chromatography.

[a]Monitor.

Protocol 17. Specific method for preparing antibody fragments from IgG_1 antibodies (esp mouse monoclonals)

Method

1. Preactivate papain by incubating at 37 °C for 30 min in 0.1 M acetate buffer pH 5.5, containing 3 mM EDTA and 50 mM cysteine.
2. Remove cysteine by column chromatography on Sephadex G25 equilibrated with 0.1 M acetate buffer pH 5.5 containing 3 mM EDTA.
3. Add pre-activated papain (ratio 1:20) to IgG_1 (10 mg/ml) in 0.1 M acetate buffer pH 5.5 containing 3 mM EDTA and react at 37 °C for 18 h,[a] adding a further aliquot of pre-activated papain after 9 h.
4. Separate fragments following elution with linear salt gradient from DEAE ion-exchange column: $F(ab)_2$ and Fab elute before undigested IgG_1 and then Fc fragments.

[a]Monitor digestion using immunoelectrophoresis.

3.4 Selection of label

There are many types of label used in immunodiagnostic tests and their use depends on a multitude of factors. For this reason it is only practical to give general guidance. The choice of label technology is also affected by the choice of assay format. For example enzyme labels are commonly used with solid-phase antibody microtitre plate (i.e. the 'Elisa' method).

A functional range of label-detection systems, bearing in mind availability of technical information and practicality in terms of general laboratory procedures, is as follows:

- light scattering technique; nephelometry and turbidimetry (Chapter 5)
- particle enhanced turbidimetry (Chapter 5)
- radioisotopes (Chapter 1)
- enzymes with colorimetric detection (Chapter 3)
- enzymes with fluorescent or chemiluminescent detection (Chapter 3)
- fluorophores with time-resolved emission (Chapter 4)
- enzymes with amplification substance cycling (Chapter 6)

These have been listed in terms of increasing sensitivity, and as a rough guide, span a target concentration from μmol $litre^{-1}$ to fmol $litre^{-1}$.

For many purposes the sensitivity of detection is not the crucial issue and choice will be dictated by existing laboratory experience and expertise, available equipment, and cost (both reagent and equipment).

For comparable sensitivity, the cost of radioisotopically labelled immunodiagnostic tests is considerably lower than other label systems, in terms of both equipment and reagents. ^{125}Iodine is the choice for radioisotope and accounts for the vast majority of radiolabelled assays. There are some ^{3}H-(tritium) labelled systems using commercially available tracers and liquid scintillation for detection. ^{57}Co has been used as part of a 'dual' label system and is only available in commercial kits for a few analytes. Specific aspects of radioisotopes and their use in immunoassays are covered in several reviews (27–29). The attributes of the other tracers are discussed in their relevant chapters.

3.5 Preparation of ^{125}I-radioiodinated tracers

^{125}I-Radioiodine is supplied as high specific activity sodium ^{125}I-iodide (carrier free) from several commercial sources, e.g. Amersham and ICN. When oxidized to radioiodine it is readily incorporated into many molecules. The simplicity of procedures and adaptability to a wide range of analyses make this a most flexible and practical system for general use. Radioisotope procedures are carried out in appropriate fume cupboards and in compliance with the necessary regulations (27). Radiolabelling of most antigens can be achieved by one of the following oxidation methods:

- lactoperoxidase, usually coupled to a solid-phase
- chloramine T
- iodogen

Using lactoperoxidase is considered to give the most mild oxidation and thereby greatly reduces the possibility of damaging the tracer molecule. Nonetheless, the simpler techniques using chloramine T or iodogen have been used successfully to prepare many and perhaps the most commonly used radiolabels. The protocols given have been optimized to yield a product with approximately 1 atom of radioiodine per molecule of tracer.

Protocol 18. Radioiodination by solid-phase lactoperoxidase (30)

Equipment

- Radioiodination facility (see appendix)
- Microvolume pipettes
- Surgical gloves
- Vortex mixer
- Stainless steel tray

Method

1. Add 10 μl 0.5 M phosphate buffer (pH 7.4) to 0.5–1.0 nmol antigen in polypropylene microfuge tube (fitted lid).
2. Add 10 μl ^{125}I-sodium iodide—37 MBq (1 mCi) (carrier free, Code No. IMS 30 from Amersham).
3. Add 10 μl solid-phase lactoperoxidase suspension (containing approximately 0.4 mg lactoperoxidase).
4. Add 5 μl hydrogen peroxide solution (diluted 1:100000 from 100 vol. in distilled water) and vortex mix.
5. After 10 min add a further 5 μl hydrogen peroxide solution and vortex mix.
6. After a total of 30 min the reaction is stopped by the addition of 100 μl 0.05 M phosphate buffer (pH 7.4) containing 0.1% sodium azide.
7. Add 100 μl 0.05 M phosphate buffer (pH 7.4) containing 1% bovine serum albumin and 1% potassium iodide to act as carrier for purification.

Protocol 19. Radioiodination by chloramine T (30)

Equipment and reagents

- Radioiodination facility (see appendix)
- Microvolume pipettes
- Surgical gloves
- Vortex mixer
- Stainless steel tray

Method

1. Add 10 μl 0.25 M phosphate buffer (pH 7.4) to 0.5–1.0 nmol antigen in polypropylene microfuge tube (fitted lid).
2. Add 10 μl ^{125}I-sodium iodide—37 MBq (1 mCi) (carrier free, IMS 30, Amersham).
3. Add 10 μg chloramine T (*N*-chloro-*p*-toluenesulphonamide from Merck) in 10 μl 0.05 M phosphate buffer (pH 7.4) and vortex mix for 30–60 s.
4. Add 10 μg sodium metabisulphite in 10 μl distilled water.
5. Add 100 μl of 0.05 M phosphate buffer containing 1% BSA and 1% potassium iodide to act as carrier for purification.

Protocol 20. Radioiodination by iodogen (30)

Equipment and reagents

- Radioiodination facility (see appendix)
- Microvolume pipettes
- Surgical gloves
- Vortex mixer
- Stainless steel tray

Method

1. Add 50 μl of iodogen (1,3,4,6-tetrachloro-3*a*,6*a*-diphenylglycoluril from Pierce Chemicals Ltd) in dichloromethane to polypropylene microfuge tube (fitted lid) and evaporate to dryness.
2. Add 10 μl 0.5 M phosphate buffer (pH 7.4) followed by 0.5–1.0 nmol antigen in 10 μl 0.05 M phosphate buffer (pH 7.4).
3. Add 10 μl ^{125}I-sodium iodide 37 MBq (1 mCi) (carrier free, IMS 30 from Amersham) and vortex mix for 30–60 s.
4. Add 100 μl 0.05 M phosphate buffer (pH 7.4) and transfer contents for purification.

Occasionally direct radiolabelling of the antigen by one of these oxidation methods will be ineffective or undesirable, usually for one of the following reasons:

(1) the tracer molecule is damaged or inferior because of susceptibility to oxidation;

(2) the radio-iodine is incorporated into a part of the molecule (epitope) directly involved in binding to the antibody;

(3) direct labelling is not possible because the tracer molecule lacks an appropriate entity, e.g. a peptide without tyrosine.

In such rare cases, the difficulty can usually be resolved by adopting some form of conjugation radiolabelling. This means the covalent coupling of a chemically reactive group already incorporating a radioiodine to the tracer molecule.

Protocol 21. Conjugation radiolabelling using ^{125}I-Bolton and Hunter reagent (31)

Equipment and reagents

- Bolton and Hunter reagent *N*-succinimidyl 3-(4-hydroxy-5-[^{125}I]-iodophenyl) propionate available commercially, e.g.. Amersham, DuPont-NEN, at specific activity ~2000 Ci/mmol (~74 Tbq/mmol)
- Radioiodination facility (see appendix)
- Microvolume pipettes
- Surgical gloves
- Vortex mixer
- Stainless steel tray

Method

1. Aliquot 1 mCi (approximately 0.5 ml) Bolton and Hunter reagent in benzene into vial.
2. Evaporate to dryness in gentle stream of nitrogen. NB. Use appropriate fume hood.
3. Add 5 nmol antigen (molar ratio 10:1 antigen to ester) in 0.1 M borate buffer pH 8.5.
4. React with mixing or gentle agitation for 15 min at 4°C.
5. Add 0.5 ml 0.2 M glycine in 0.1 M borate buffer, pH 8.5, for 5 min at 4°C to react unchanged ester.
6. Purify.

NB. *N*-Succinimidyl [2,3–^{3}H] propionate may be used to label with tritium in a similar manner.

Conjugation radiolabelling usually gives products with lower specific activity than found by direct labelling.

3.6 Purification of ^{125}I-radioiodinated tracers

The products from radioiodination require purification before use. Any impurity in the original material could be radioiodinated and would require removal. It is also possible that radioiodination would generate additional impurities, also radioiodinated, following damage and degradation from oxidation. For this reason, full resolution of products during purification is desirable and gives more reliable radiolabelled tracers.

A suitable purification method for any given individual tracer will be selected on the basis of physico-chemical characteristics. ^{125}I-labelled steroids and drugs are often purified by HPLC (30); peptides by gel exclusion chromatography (30) or ODS adsorption (32); and proteins by gel exclusion chromatography (33). Other techniques for purifying radiolabelled tracers include TLC (34), gel electrophoresis (35) and micro-immunosorbent (36).

A general method can be given for purifying radioiodinated antibodies, usually monoclonals or affinity-purified polyclonals (see *Protocol 22*).

Protocol 22. Purification of ^{125}I-antibodies

Equipment and reagents

- 30 × 1 cm chromatography column (e.g. Pharmacia)
- Sepharyl S300 HR (Pharmacia)
- Peristaltic pump with flow rate 3–6 ml per hour, e.g. Watson–Marlow 302F/RL with 308MC/A head and appropriate tubing (orange/green)
- Small gamma counter, e.g. Mini-assay 6.20 (Mini-Instruments Ltd)
- Fraction collector
- Elution buffer (phosphate 0.05 M, pH 7.4) containing 1% BSA and 0.1% sodium azide

Method

1. Fill column with Sepharyl S300 HR and equilibrate with elution buffer at 3 ml per hour.
2. Add 100–200 μl of elution buffer containing potassium iodide (1%) as carrier to radioiodination vial and gently mix.
3. Add 100 μl sucrose solution (saturated) and mix gently.
4. Load contents into the column from the end of column supply tube, by dipping it to the bottom of the vial.
5. Remove when loading complete, and transfer to supply of elution buffer.
6. Collect fractions of 1 ml and continue to elute column at 3 ml per hour overnight.
7. Count radioactivity in fractions by positioning in holder some 10 cm above well. Cover well to protect from contamination.
8. Plot profile and identify purified product (see *Figure 5*).

For storage, purified tracers should be diluted to between 10–20 μg/ml with buffer (e.g. phosphate 0.05 M, pH 7.4) containing 1% protein (e.g. bovine serum albumin). Aliquots are stored frozen or lyophilized and left at 4°C. Mannitol (2%) is added to the diluent buffer if aliquots are to be lyophilized.

3.7 Selection and preparation of standard or calibrant

A successful application of any immunodiagnostic test is intimately related to the type and accuracy of calibration or standardization. It is also useful to distinguish between primary standards, sometimes referred to as calibrants,

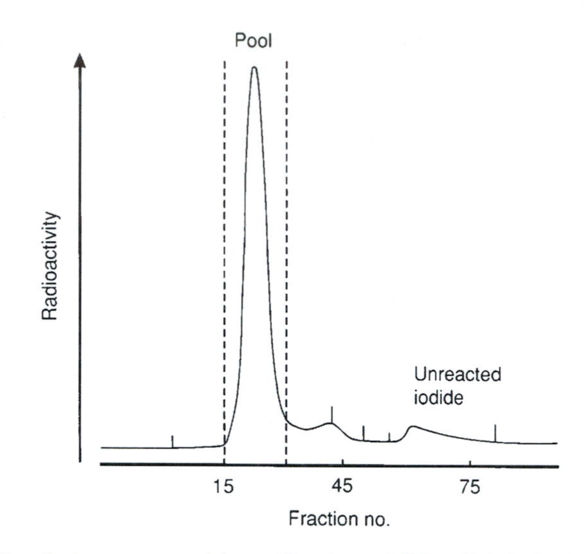

Figure 5. Profile of chromatographic purification of ^{125}I-radiolabelled monoclonal antibody; this typical example illustrates separation of the purified product from a small peak of contaminant and unreacted radioactive iodide. Fractions containing the product are pooled, diluted and aliquoted for storage and use.

and secondary standards, often called 'working' standards. Useful information on preparation of standards is available in some reviews (37,38).

Many common or widely used immunodiagnostic tests benefit from the availability of international reference materials which are used for calibration. These reference materials are prepared under the auspices of international, regional and national organisations, such as the World Health Organisation (37). Their use has led to improved compatibility of results from different laboratories. This harmonization is an essential step in the full accretion of diagnostic experience and interpretation. It is particularly important because individual workers or laboratories are rarely able to command the complete picture necessary to formulate an all-inclusive diagnostic protocol.

It is difficult to give precise and detailed guidance on selecting the appropriate standard material because of the vast range of possible analyses. However, the basic aspects of standardization are generally similar. In the first place a distinction needs to be made between the standard itself and an appropriate matrix. The simple rules that apply are as follows:

1. The standard material and analyte in samples should be identical, or at best as similar as practicably possible.
2. Any degree of heterogeneity in the analyte should be reflected in the standard preparation, e.g. same proportion of different forms.
3. The substance used as standard should be chemically pure.

4. The matrix, e.g. serum, urine, etc., should be similar for both standards and sample.
5. The assigned concentration for a standard preparation must be confirmed by a similar test, i.e. reference immunoassay and not a bioassay. Some preparations may give quite different potencies in bioassays compared to immunoassays.
6. Standards must be produced in a form which ensures maximum stability.

The chemical identity and degree of purity of the material selected for use as standard has to remain a matter of judgement. However, it remains a useful working rule that the degree of purity should be the highest possible in the first instance. It may prove practical to accommodate a less pure material once calibration has been established and it is possible to make significant comparisons.

The general approach to selecting a suitable matrix is to use something similar but which is lacking any endogenous analyte. This means using one of the following:

1. 'Synthetic' matrix, involving a formulation to mimic the main components of the matrix using purified materials, e.g. for 'synthetic' serum

serum albumin (bovine)	4–10%
gamma globulin (bovine)	0.5–1%
hydrolysed gelatin	0.5–1%
mannitol	0.5–1%

   in phosphate buffered saline (0.05 M (pH 7.4) phosphate with 0.15 M sodium chloride).
2. Matrix from different species, e.g. horse serum can be substituted for human serum where the comparative molecule in the horse is immunologically distinct and does not cross-react in the test. This applies for many peptide or protein analytes.
3. Matrix from the same species where the endogenous analyte is removed by adsorption. This could be by the use of specific immunosorbents or by simple adsorption using charcoal (see *Protocol 23*).

Protocol 23. Removal of endogenous analytes from serum using a charcoal/cellulose column

Equipment and reagents

- Activated charcoal, acid washed
- Microcrystalline cellulose, e.g. Sigmacell 20
- Filter discs, e.g. GF/C, Whatman
- Columns, e.g. 50 ml disposable syringes (e.g. B.D. Plastipak)
- 100 ml disposable plastic screw-capped containers
- Stands and clamps
- Distilled water

Protocol 23. *Continued*

Method

1. For each column, weigh 8 g charcoal together with 2 g cellulose into a container, cap and shake vigorously to dry mix the contents. The mixture should acquire 'free flowing' properties. Leave to settle.
2. Mix 10 g of cellulose with 40 ml of water and mix to form a thick slurry (sufficient for approximately 10 columns).
3. Fix the syringe barrels upright in a clamp and place a 2.5 cm filter (GF/C) disc flat in the bottom of each column.
4. Moisten with a few drops of water so that the filter adheres to the base of the syringe barrel.
5. Using a Pasteur pipette, add a few ml of the slurry into the column and allow the water to drain through. Continue adding the cellulose to the columns to form a layer about 1 cm deep. A small amount of cellulose may leak through; however, if cellulose immediately flows past the filter, then abandon and start again.
6. Allow excess liquid to drain from the cellulose layer.
7. Add 25 ml of water to each of the charcoal–cellulose mixtures, recap the container and shake vigorously to form a slurry. This slurry should not 'stick' to the container walls, if it does, add an extra few ml of water and mix again. Do not add more than 30 ml of water in all.
8. Add a few ml of the charcoal–cellulose slurry using a Pasteur pipette so as not to disturb the cellulose layer then the remainder can be poured into the column. Leave until the mixture has settled and excess liquid has drained through. This will take about one hour.
9. Add filtered serum to column and allow to flow through into pot. Continue to add serum up to 100 ml per column. Displace any serum retained on the column using an additional volume of water.

4. Matrix with no endogenous analyte serum from a normal subject where the analyte is only found in the pathological state.

The material is stabilized by use of preservative, e.g. 0.1% sodium azide, filtering to remove any contaminating microorganisms, i.e. 0.2 micron filters, and subsequent storage, either frozen at –20°C or lyophilized and kept at 4°C. Aliquoting standards is important to avoid repeat freeze/thaw cycles, which can often be damaging.

4. Assay design

Studies on the theoretical basis of immunoassays have made significant contributions to our understanding of the principle of assay design (21–23).

These are primarily of importance in providing general guidelines and in correcting some of the mistaken arbitrary rules followed by many assayists (24). Nonetheless, much of the design process is influenced by other practical issues such as familiarity of technique, availability of appropriate equipment, experience in practical application and cost. It is common for most assay development to be based on an empirical approach. The principal steps in assay design are:

- choice of assay format including method of separation
- selection of general reagent components
- sample preparation
- optimal protocol and concentration of reagents
- validation of assay performance

4.1 Assay format

After selecting or preparing the three key reagents, i.e. antibody, tracer with appropriate detection system and a suitable standard (see Section 3), it is necessary to decide which format to use. There are three basic and distinct types of assay as follows:

(1) limited antibody reagent concentration (competitive type)—IA;
(2) excess antibody reagent concentration (non-competitive, using one specific antibody)—IMA;
(3) excess antibody reagent concentration (non-competitive, using two specific antibodies)—IMA.

These basic types are illustrated diagrammatically in *Figures 6*, *7* and *8*, respectively. The factors affecting the choice of basic format are given in *Table 4*. An additional aspect of the format is the technique for separating free and bound fractions.

Over the years many methods have been used to separate the fractions (20). Whilst many have been non-specific, but nonetheless adequate, most procedures currently used are specific. The majority of tests now use solid-phase reagents, i.e. antibody or antigen linked to particulate material or the surface of the reaction vessel itself. The latter has become the most widespread due to its effect on simplifying work procedures and the potential for enhancing assay performance. This subject is detailed in Chapter 2. Solid-phase reagents are almost mandatory in both IMA formats and can be very useful in IA particularly when used in conjunction with a second antibody.

A useful substitute for solid-phase reagents, whilst still retaining specificity of separation, uses a second antibody in the liquid phase. A specific anti-species antibody, e.g. anti-rabbit immunoglobulin, is used in appropriate dilution to form a precipitable complex with the reagent antibody, i.e. in this

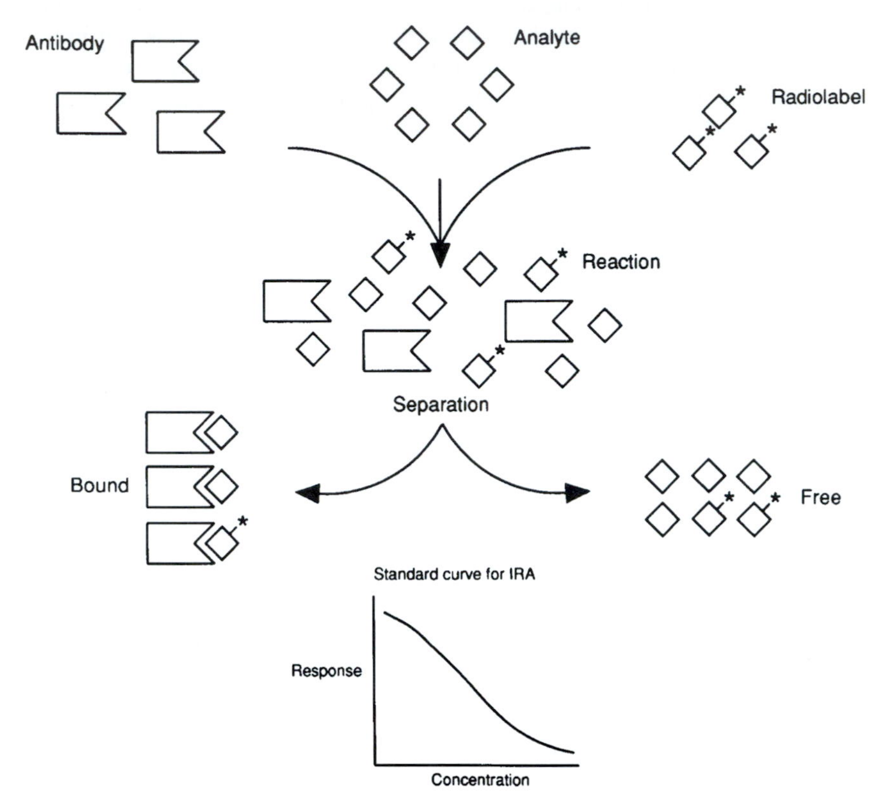

Figure 6. Diagrammatic representation of immunoassay (IA, i.e. competitive) using radiolabel; an IA uses a limited concentration of antibody in proportion to the amount of analyte. Assays using other labels, e.g. enzyme, are based on the same principle.

case a rabbit anti-analyte antibody. The ability to precipitate the complex is enhanced by the use of polyethylene glycol (average polymer size 6000) set between 2 to 4% w/v final concentration. The second antibody can be added at any stage of the primary reaction, i.e. at the beginning or towards the end as a second stage reaction. The use of carrier serum, i.e. non-immune ('normal') serum from the same animal species used for the primary antibody, is recommended to provide adequate precipitable material. This normal serum would generally be added at the beginning of the reaction, together with the primary antibody.

Both these reagents need to be titrated to find the optimal dilutions for maximum precipitation. Either too much or too little would lead to suboptimal precipitation (39). A simple example is given in *Protocol 24* with typical concentrations.

Table 4. Decisive factors for selecting assay format

<table>
<tr><th>Assay format</th><th>Immunoassay</th><th colspan="2">Immunometric assay</th></tr>
<tr><th></th><th>IA</th><th>IMA</th><th>IMA (two-site)</th></tr>
<tr><td>Requirements</td><td>Purified antigen for tracer</td><td colspan="2">Purified antibody for tracer</td></tr>
<tr><td></td><td></td><td></td><td>Antiserum for solid-phase (polyclonal or monoclonal—need not be purified)</td></tr>
<tr><td>Separation</td><td>(i) Ab_2/4% PEG with centrifugation
(ii) Solid-phase Ab_2</td><td>Solid-phase antigen</td><td>Solid-phase antibody</td></tr>
<tr><td>Advantages</td><td>No 'hook' effect</td><td colspan="2">Potentially more sensitive</td></tr>
<tr><td></td><td></td><td></td><td>More specific</td></tr>
<tr><td></td><td></td><td colspan="2">Shorter reaction time</td></tr>
<tr><td></td><td></td><td colspan="2">Wider working range (approx. 1000-fold)</td></tr>
<tr><td>Disadvantages</td><td>Less sensitive
Narrow working range (approx. 100-fold)</td><td></td><td>Potential hook effect</td></tr>
</table>

Ab_2 = second antibody.

Protocol 24. The use of second antibody and carrier serum in PEG assisted method—selecting optimal concentrations

Method

1. Set up assay tubes containing 'zero' standard, first antibody at dilution used in assay, tracer, and assay buffer, and react for the time indicated in the assay protocol (total volume approx. 500 μl). Tubes with all reagents except for first antibody are used for measuring NSB.
2. Add 50 μl of second antibody at dilutions of 1:5, 1:10, 1:20, 1:40, 1:80 or 1:160 to tubes in duplicate.
3. Vortex mix and add 50 μl of carrier serum at dilutions of 1:50, 1:100 and 1:200 for NRS (1:500, 1:1000 and 1:1500 for normal sheep serum).
4. Leave at ambient temperature for 1–2 h.
5. Add 1 ml 4% PEG containing 0.1% Triton X-100 and vortex mix thoroughly.
6. Centrifuge at 200 g at ambient temperature for 30 min.
7. Decant into sink and leave to drain on absorbent paper for 10 min before counting.

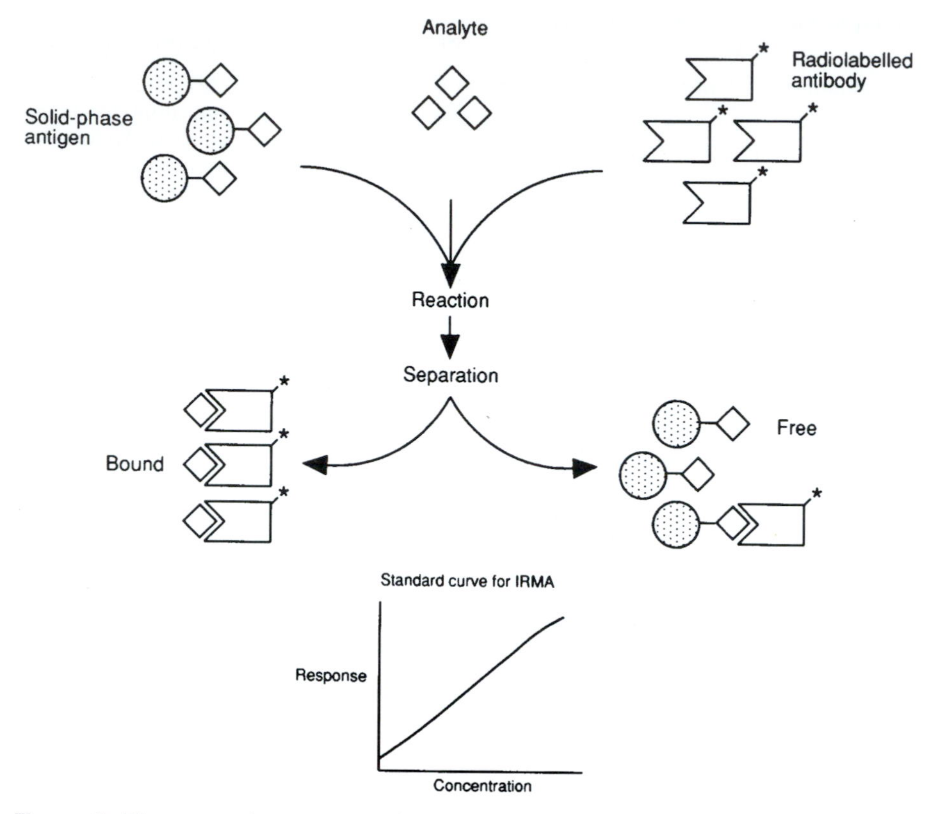

Figure 7. Diagrammatic representation of an immunometric assay (IMA, i.e. non-competitive) using a radiolabelled antibody; an IMA uses the antibody reagent in relative excess. Assays using other labels, e.g. enzyme, are based on the same principle.

4.2 General reagents and additives

In general the critical reagents, i.e. antibody, analyte and labelled tracer, are allowed to react in a buffered solution containing various components or additives, each one used for a particular purpose. It is common to use 'universal' buffers with appropriate additives. Particular buffers or additives are occasionally necessary when used in connection with particular systems. These will be discussed in the appropriate chapters.

The commonly used additives or components are given below:

- non-specific protein carriers, e.g. serum albumin
- physiological salt concentration, e.g. sodium chloride
- specific proteins for eliminating 'heterophilic' antibody effect, e.g. gamma globulins

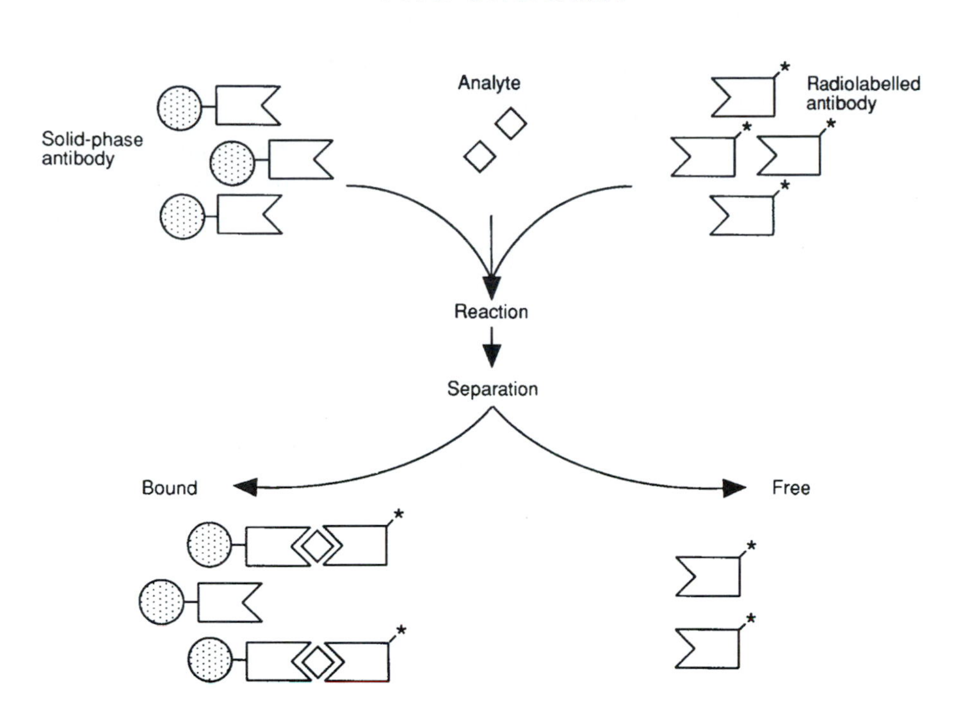

Figure 8. Diagrammatic representation of two-site immunometric assay using a radiolabelled antibody; these assays use two antibodies, one on the solid phase and the other as labelled. The standard curve is similar in characteristics to *Figure 7*.

- chemicals to eliminate binding protein effect, e.g. anilino-sulphonic acid (ANS)
- detergents to reduce non-specific binding especially with solid-phase, e.g. tween 20
- preservatives
- EDTA to reduce calcium ion interference in second antibody reaction

The use of these components will depend upon particular applications. Guidelines for their use are summarized in *Table 5*.

4.3 Sample preparation

One of the main attributes of immunodiagnostic tests is specificity. This means that they can be applied to complex mixtures such as biological fluids like serum or urine. In general, it is unnecessary to modify or pretreat samples. However, there are examples of sample preparation which can enhance assay performance as follows:

(1) concentration of analyte where assay has insufficient sensitivity;
(2) extraction of analyte to enhance specificity or reduce matrix effect;
(3) extraction of interfering species to enhance specificity;
(4) precipitation of binding proteins.

In general organic solvents have often been used to extract and concentrate small non-polar molecules like steroids or drugs. This procedure can be particularly useful when applied to samples like urine, water supplies, or agricultural and industrial liquid wastes. The extracted material can be easily concentrated following the evaporation of the solvent and then by reconstituting in a much smaller volume of buffer. Separation of the organic and aqueous phase can be simplified by freezing. The organic phase can normally be decanted leaving the aqueous phase as a frozen plug in the tube.

The use of specific absorbents such as ODS–silica in small prepackaged capsules is a very convenient way to prepare samples for analysis. When used in conjunction with appropriate manifolds the method can be scaled up to

Table 5.

Additive	**Example**	**Reason**
Protein	Bovine serum albumin 0.1–0.5% (w/v); gelatin 0.1–0.2% (w/v); serum 1–5% (v/v)	Reduce non-specific binding of labelled tracer to various surfaces
Detergent	Triton X-100 0.01–0.1% (v/v); Tween 20 0.05–0.5% (v/v)	Reduce non-specific binding to solid-phase reagents
Protease inhibitors	Trasylol (aprotinin); Bacitaracin	Eliminate degradation of certain components, especially the tracer, sensitive to proteolytic activity usually originating with the specimen
Specific blocking agents	Salicilate; ANS (8-anilino-l-naphthalene sulphonic acid)	Eliminate binding of either analyte or labelled analyte to specific binding proteins
Specific proteins or serum	Gamma globulins; mouse serum	Remove effect of heterophilic antibodies or specific antiglobulin antibodies
Preservatives	Sodium azide 0.05–0.1% (w/v); Thiomersal 0.02% (w/v); Bronidox 0.1% (v/v)	Useful to prolong shelf life of buffers and prevent growth of microorganisms
Others	EDTA 0.01 M; Heparin (10 IU/tube)	Optimize second antibody reaction (see Chapter 4)
	PEG (Polyethyleneglycol 6000) 1.4% (w/v)	Enhance visual precipitation especially in turbidimetry

NB. Typical buffer—Phosphate buffered saline 0.05 M phosphate, pH 7.4 containing 0.15 M sodium chloride.

process a large number of samples. These techniques have been used extensively for monitoring agrochemical or environmental pollution, investigating the presence of a variety of organic residues in complex matrices like liver tissue in forensic diagnoses, screening for drugs of abuse in both humans and animals, and in monitoring metabolites in pharmaceutical studies. Protocols are available from commercial suppliers of cartridges.

Endogenous binding proteins in samples can cause difficulties in estimating accurate concentrations of some analytes. This effect can be resolved by simply denaturing and precipitating the proteins providing that the analyte is not affected. Reduction of pH, e.g. using formic acid, or addition of chaotropes, e.g. 3 M urea, have been useful.

4.4. Optimal protocol and concentration of reagents

Development and optimization of the assay at this stage, i.e. after selecting the key immunoreagents, detection system and basic assay format, will be a matter of ensuring that the potential performance is realized. Potential sensitivity and specificity are intrinsic aspects of the particular antibody selected as discussed earlier, and cannot be changed without changing the antibody. These parameters will be enhanced or compromized by subsequent selections of assay detection and format. Potential performance characteristics may be further altered following necessary or desirable constraints on reaction time, working patterns, temperature and volume.

This final stage of development is achieved by empirical 'titration' of the key reagents. Thus using a standard protocol the concentration of each reagent in turn is 'titrated' to achieve optimal performance, whilst keeping the concentration of other reagents constant. In most cases, the initial concentration is selected, usually on the basis of estimation, to give an excessive response. This is then subject to various dilutions, (successive doubling dilutions are often most satisfactory), and then an equal sample of all dilutions is tested individually under identical circumstances.

A list of typical characteristics and possible modifications for each basic assay type is given below to assist in initial selections. The final formulation is made only after consideration of a particular application.

Typical characteristics for RIA (^{125}I-tracer):

- *^{125}I-Tracer: specific activity approx. 2000 μCi/nmole or 74 Mbq/nmole (NB. using 'carrier free' ^{125}iodide); 20–30000 cpm per tube*
- *Antibody: final dilution to give approximately 20–50% binding of tracer*
- *Sample: 10–100 μl*
- *Total volume: 250–500 μl*
- *Separation: solid phase second antibody or second antibody/ 4% PEG pre-*

cipitation. Modifications adjusted to give maximum binding but minimum NSB.

- *Decrease concentration of antibody to increase sensitivity and precision. This may necessitate increasing time of reaction.*
- *Varying sample size should change sensitivity proportionately. NB. Increase in sample size may introduce matrix bias.*
- *Increase in the specific activity of the ^{125}I-tracer may improve sensitivity but only minimally (possibly up to 10%); NB. could lead to damage or loss of stability of tracer.*
- *Delayed addition of ^{125}I-tracer may improve sensitivity by a factor of two. the delay in adding tracer can be varied, but 50% of reaction time can be used as an initial test.*
- *Reducing NSB by washing precipitated or solid-phase bound fraction$_s$ will often improve sensitivity and precision. This can be achieved by adding 1–2 ml of diluted buffer immediately prior to physical separation. Further wash-steps are often counterproductive.*
- *Changes to specificity are usually improved by some form of sample purification (see previous section). A change in the type of tracer has been known to affect specificity but not in a predictable manner.*

Typical characteristics for IRMA (^{125}I-tracer):

- *^{125}I-antibody specific activity approx. 1500 μCi/nmole (= 10 μCi/μg) 55 Mbq/nmole (NB. using 'carrier free' ^{125}iodide); 50–100 000 cpm per tube.*
- *Solid-phase antibody (if two-site) to give maximum achievable binding with minimum NSB, possibly as low as 0.2%. A minimal NSB value is more important than a small increase in binding.*
- *Other features similar to RIA*

Modifications

- *Successive washing steps, i.e. addition of 1–2 ml of dilute buffer containing detergent, will increase sensitivity and precision (up to a maximum of three washes).*
- *Increase in specification activity of tracer (labelled antibody) will decrease counting error and enhance sensitivity to same extent. This could lead to damage or loss of stability of tracer.*
- *Reaction of solid-phase antibody with sample and subsequent reaction with labelled antibody carried out as two separate stages may remove matrix effect*

and enhance sensitivity. This two-step format also reduces or eliminates any 'hook' effect.

In recent years attention has been focusing on the length of time that an assay takes. Manipulating conditions to shorten the time is not difficult; however, doing so often compromises other aspects. It is therefore necessary to monitor this carefully. The following points are useful to consider:

1. The time is best adjusted to suit particular working patterns.
2. An increase in operating temperature will speed up the reaction.
3. An increase in reagent concentration will improve the rate of reaction, but may lead to an increase in NSB, which will compromise sensitivity and precision, or reduce specificity.
4. The use of a shaker or rotator to mix components during the reaction will increase the rate for solid-phase reagents.
5. It is not strictly necessary to allow the reaction to reach equilibrium before measurement, in fact many assays are stopped much earlier at a non-equilibrium state. This does increase the potential for assay drift. With some systems it is possible to measure kinetic rates rather than end-points for earlier measurement and thus shorten assay time.

4.5 Assay validation

The final stage of development is validation of performance characteristics. Although this may vary for different applications, emphasis should be given to the following tests:

- intra-assay and inter-assay precision
- drift
- comparison with reference method
- recovery and dilution tests
- response to interfering factors

4.5.1 Intra-assay and inter-assay precision

These are best assessed by the use of precision profiles (see *Protocol 25* and *Figure 9*). Precision profiles are only accurate when calculated from an adequate number of reagents over the working range. This will generally involve a minimum of 25 samples with values that span the range.

Protocol 25. Calculation for intra-assay precision profile

Method

1. Calculate the mean response (y) and standard deviation in response (SD) for each set of replicates.
2. Plot the mean response (y) against concentration [analyte] for the calibrators. This is the calibration curve.
3. Square the standard deviation in response for each set of replicates to obtain the variance in response.
4. Divide the concentration range of the assay into 5–10 'bins' (smaller concentration ranges). Sum the variances of response in each 'bin' and average. The 'bin' will be represented by the single average variance value.
5. Plot the average 'bin' variance against the average 'bin' concentration. This is the response error relationship (RER) error profile.
6. Use least squares regression to fit one of two models to data, variance $= a + by + cy^2$ or variance $= a(y)^b$.
7. Calculate the modelled variance for each bin from the curve in (6) above, calculate the square root of the variance to obtain the standard deviation (SD).
8. Calculate the gradient of the calibration curve at concentration levels corresponding to each mean 'bin' concentration.
9. Use the gradient of the calibration curve to transform the SD in response for each 'bin' to concentration errors.

   $$\text{Error of concentration} = \frac{\text{SD of response}}{\text{Slope of response curve at this point}}$$

10. To obtain a precision profile in terms of SD, plot the concentration error against concentration.
11. To obtain a precision profile in terms of % coefficient of variation (%CV), divide this concentration error by the average concentration value of the 'bin' and multiply the result by 100 to obtain %CV. Plot %CV against concentration, as in *Figure 9*.

4.5.2 Drift

If a sample shows significantly different results depending on where it appears in the assay sequence, the phenomenon is commonly referred to as drift. This may be a problem with high-throughput assays, particularly if the reaction time is short, i.e. non-equilibrium. An assay can be tested for potential drift by

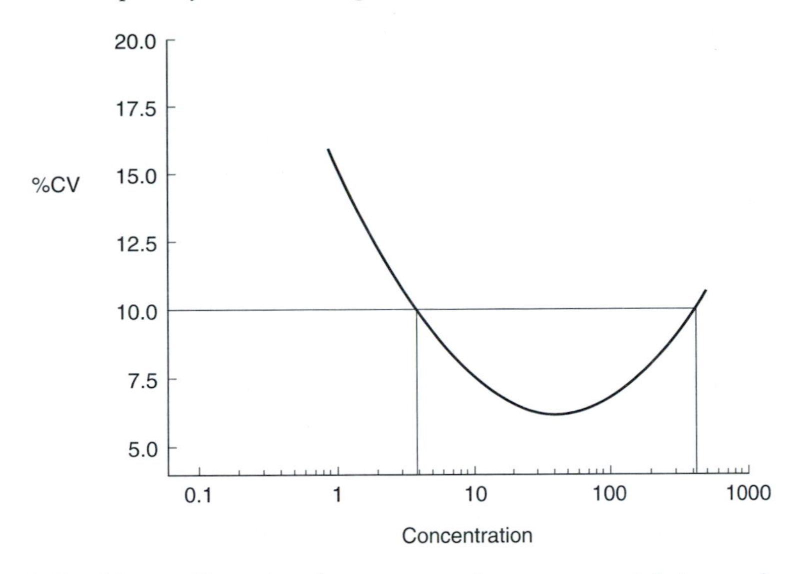

Figure 9. Precision profile; a plot of average error in measurement, in terms of concentration, against the concentration of analyte. In this example an estimated working range is indicated by defining an acceptable error as 10% CV.

repeated analysis of a suitable pool or pools at successive positions. Any difference must be tested for statistical significance.

4.5.3 Comparison with a reference method

If a reference method is described and available, it is important to compare results from a batch of samples (at least 50) analysed by both it and the developed assay. These results should be analysed by linear regression analysis. A suitable reference method is usually one involving definitive analytical steps, e.g. physio-chemical methods coupling chromatographic purification with specific detection.

It may also be possible to compare results with target values defined by analyses from a significant number of other laboratories, i.e. all laboratory trimmed means (ALTM) or method means for pooled source material distributed in external quality assessment schemes (EQAS)—see Chapter 8.

4.5.4 Recovery and dilution tests

Recovery: Measurements are made on samples before and after the addition of a known amount of analyte. The difference in the values is expressed as a percentage of the added mass. The concentration of analyte added is sufficiently high to minimize any volume change; it is preferable to correct data for small volume changes. This test can be applied to any sample no matter what the level of endogenous analytes, as long as the final result is within working range. A satisfactory result is not significantly different (statistically) from 100%.

Dilution: Samples are prepared at various dilutions, e.g. neat (undiluted), 1:2, 1:4, etc., in selected matrices. Following measurement, results should be linear, i.e. dilute in parallel to the potency of the calibrant assessed in the same way.

Recovery and dilution experiments assess both accuracy of calibration and appropriateness of matrix. However, it is difficult to discriminate between these two if the results are deficient.

4.5.5 Response to interfering factors

Inappropriate responses to interfering factors or cross-reactants are difficult to assess in the absence of a totally reliable reference method. Measurements can be monitored for a change after the addition of suspected interfering factor or known cross-reactants.

It is sometimes possible to use regression analysis for groups of specimens that appear to be similar in all respects other than the presence or absence of interfering factors.

Typical interfering factors vary, depending on methods and applications. Some examples of interference in clinical samples are:

- sample specific effects, e.g. degradation due to haemoglobin, presence of lipids etc.
- effects of storage, e.g. time and temperature
- heterophilic antibodies
- rheumatoid factors
- antibodies reacting with label, e.g. anti-peroxidases or specific anti-mouse antibodies (present after patient treated with mouse monoclonal antibodies)
- abnormal proteins, especially binding proteins
- high salt concentrations, e.g. urine
- chaotrophic components, e.g. urea

Appendix

Radioiodination facility

All work using radioisotopes is usually subject to various regulations, e.g. The Ionising Radiations Regulations 1985 (Her Majesty's Stationary Office 1985) and subsequent Approved Code of Practice (HMSO) and Guidelines to the IRR (HMSO). No work should be carried out without adequate training and supervision. All radioiodinations using more than 50 μCi (1.85 MBq) of ^{125}I-sodium iodide should be carried out in a special radioiodination facility, see *Figure 10*. The operation is carried out in a fume cupboard operating in a specified controlled area. A suitable facility is illustrated in the figure and notes on its design follow.

Surfaces

All surfaces should be accessible and easy to clean. Surfaces should be non-adsorptive and generally resistant to chemical attack, heat, and fire. The flooring should be a continuous surface (PVC is suitable) with welds, where necessary, of at least 50% PVC. High gloss paint is used on walls and ceilings and light gloss-epoxide for woodwork. Exposed pipe work on walls and ceilings is to be kept to a minimum. Radiators and light fittings should be flush.

Benching and sinks

Sinks should be stainless steel, where possible, and the hand wash basin should have elbow (or foot) operated taps. Bench surfaces can be melamine and should be continuous with a slight raised lip at the front to retain any spillage. Alternatively, work should be carried out in trays (stainless steel, if possible) to restrict spillage or contamination. Non-adsorptive plumbing, e.g. glass or high density alkathene, should be used for drainage from main sink.

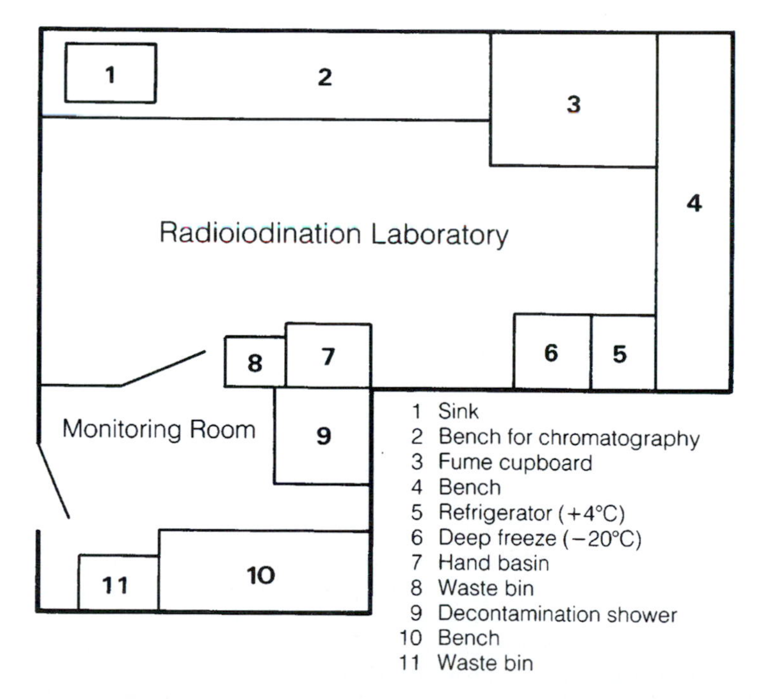

Figure 10. Radioiodination facility; diagram shows a typical layout for the preparation, purification and storage of radioiodinated traces.

Fume cupboard and ducting

The fume cupboard should be sited away from any doors or emergency exits. The fully enclosed cabinet must have an air flow of >0.5 m/s at a working aperture of about 100 × 30 cm.

The exhaust should discharge directly to open air through separate ducting which should not be in close proximity to windows or doors. Suitable fume cupboards and ducting can be supplied complete from commercial sources. The working surface in the fume cupboard should be non-adsorptive (e.g. stainless steel or melamine).

Monitoring room

A monitoring room or area is useful at the entrance to the radioiodination area. Ideally this should be separated from the radioiodination laboratory by a continuous wall with closing door (NB the wall or door will require an air-flow valve).

Air flow

The air flow should be from the corridor, through the monitoring room to the laboratory, into the fume cupboard and out through the ducting. A negative pressure in the laboratory with respect to the monitoring room of about 1/10″ to 1/4″ buffer pressure is useful during the iodination to limit the chance of contamination.

References

1. Uhlenhuth, T. (1903) *Dtsch. med. Wschr.* **29**, 39.
2. Bechhold, H. (1905) *Z. phys. Chem.* **52**, 185.
3. Landsteiner, K. (1943) *The specificity of seriological reactions.* Harvard University Press, Cambridge, Massachusetts.
4. Heidelberger, M. (1929) *J. Exp. Med.* **50**, 809.
5. Coons, A. H., Creech, H. F. and Jones, R. N. (1941) *Proc. Soc. Expt. Biol. Med.* **47**, 200.
6. Oudin, J. (1946) *C. R. Acad. Sci.* **222**, 115.
7. Ouchterlony, Ö. (1948) *Acta. Path. Microbiol. Scand.* **25**, 186.
8. Arquila, E. R. and Statvitsky, A. B. (1956) *J. Clin. Invest.* **35**, 458.
9. Ekins, R. P. (1960) *Clin. Chim. Acta.* **5**, 453.
10. Yalow, R. S. and Berson, S. A. (1960) *J. Clin. Invest.* **39**, 1157.
11. Wide, L., Bennich, H. and Johansson, S. G. O. (1967) *Lancet* **2**, 1105.
12. Miles, L. E. M. and Hales, C. N. (1968) *Nature* **219,** 186.
13. Haberman, E. (1970) *Z. Klin. Chem. Klin. Biochem.* **8**, 51.
14. Wide, L. (1971) In *Radioimmunoassay methods* (ed. K. E. Kirkham and W. M. Hunter), p. 405. Churchill Livingstone, Edinburgh.
15. Addison, G. M. and Hales, C. N. (1971) *Horm. Metab. Res.* **3**, 59.
16. Kohler, G. and Milstein, C. (1975) *Nature* **256**, 495.

17. Bangham, D. R. (1983) In *Immunoassays for clinical chemistry,* 2nd edn (ed. W. M. Hunter and J. E. T. Corrie), p. 27. Churchill Livingstone, Edinburgh.
18. Mancini, G., Nash, D. R. and Heremans, J. F. (1970) *Immunochemistry* **7,** 261.
19. Laurell, C.-B. (1966) *Anal. Biochem.* **15**, 45.
20. Edwards, R. (1996) *Immunoassays,* p. 26. Wiley, Chichester.
21. Ekins R. P. (1976) *General principles of hormone assay* (ed. J. H. Loraine and E. T. Bell), p. 1. Churchill Livingstone, Edinburgh.
22. Ekins R. P. (1997) In *Principles and practice of immunoassays,* 2nd edn (ed. C. P. Price and D. J. Newman), p. 173. Macmillan, London and Stockton Press, New York.
23. Jackson T. M. and Ekins R. P. (1986) *J. Immunol. Meth.* **87**, 13.
24. George, A. J. G. (1997) In *Principles and practice of immunoassays,* 2nd edn (ed. C. P. Price and D. J. Newman), p. 65. Macmillan, London and Stockton Press, New York.
25. Andersson, L. I. and Mosbach, K. (1997) In *Principles and practice of immunoassays,* 2nd edn (ed. C. P Price and D. J. Newman), p. 139. Macmillan, London and Stockton Press, New York.
26. Siddle, K. (1990) In *Peptide hormone secretion* (ed. J. C. Hutton and K. Siddle), p. 97. IRL Press, Oxford University Press, Oxford.
27. Edwards, R. (1997) In *Principles and practice of immunoassays,* 2nd edn (ed. C. P. Price and D. J. Newman), p. 325. Macmillan, London and Stockton Press, New York.
28. Hunter, W. M. and Corrie, J. E. T. (eds) (1983) *Immunoassays for clinical chemistry*. Churchill Livingstone, Edinburgh.
29. Bolton, A. E. and Hunter, W. M. (1986) In *Handbook of experimental immunology* (ed. D. M. Weir), Vol. 1, p. 26.1. Blackwell Scientific, Oxford.
30. Edwards R. (1990) In *Radioimmunoassay*, p. 71. IRL Press, Oxford.
31. Bolton, A. E. and Hunter, W. M. (1973) *Biochem. J.* **133**, 529.
32. Bennett, H. P. J., Hudson, A. M., McMartin, C. and Purdon, G. E. (1977) *Biochem. J.* **168,** 9.
33. Edwards, R., Little, J. A., Zaman, M. R., Knott, J. A. and Newman, D. J. (1992) In *Proceedings symposium, Vienna 1991,* p. 205. International Atomic Energy Agency, Vienna.
34. Thorell, J. I., Ekman, R. and Malinquist, M. (1982) *Radioimmunoassay and related procedures in medicine,* p. 147. International Atomic Energy Agency, Vienna.
35. Linde, S., Hansen, B. and Lernmark, A. (1983) *Adv. Enzymol.* **92,** 309.
36. Hales, C. N. and Woodhead, J. S. (1980) *Adv. Enzymol.* **70,** 334.
37. Stenman, U-H. (1997) In *Principles and practice of immunoassays,* 2nd edn (ed. C. P. Price and D. J. Newman), p. 243. Macmillan, London and Stockton Press, New York.

2

Solid-phase supports

STUART BLINCKO, SHEN RONGSEN, SHEN DECUN, IAIN HOWES and RAYMOND EDWARDS

1. Introduction

Optimum immunoassay performance is achieved when only the reaction of a specific antibody with its antigen is measured (1). Assuming the antibody preparation is of high affinity and high specificity then the key to achieving highly sensitive and highly specific assays is the reduction of non-specific signals. Such signals are caused by a number of effects. These include non-specific interactions of the labelled immunoreagents, background signals from the sample and assay components and detector noise (e.g. optical and electrical noise). Furthermore the method adopted must demonstrate a high degree of precision to achieve maximum sensitivity (2).

Non-specific interactions of labelled immunoreagents (i.e. labelled specific antibodies and antigens) with sample and assay components are often referred to as non-specific binding (NSB). These interactions include non-specific protein–protein associations and binding to assay cuvettes. In addition to NSB the sample itself (e.g. serum, plasma, whole blood, urine) may contribute to background. For example, most biological fluids have endogenous fluorescence which could affect the sensitivity of a fluorescence immunoassay. Assay components (e.g. buffers, reagents, cuvettes) may also contribute to the background signal in a similar manner.

A common strategy for minimizing NSB and the background due to sample and assay components involves the use of solid phase bound immunoreagents (i.e. antibody or antigen bound to materials insoluble in the assay buffer) (3,4). This enables the solid phase specifically bound component of an assay to be easily and quickly removed from the reaction medium. The solid phase may then be washed (extensively if necessary) so reducing the NSB and removing the presence of sample and assay components. Additionally it has been reported that washing can improve the specificity of an assay for small molecules (5).

Another advantage of using solid phases is that they enable assay components in multistep assays to be changed with ease. For example, a horseradish peroxidase–antibody conjugate may be incubated with a solid phase bound antigen in phosphate buffered saline (PBS) pH 7.4. Following aspiration and washing away of the unbound conjugate an enzyme substrate (such as tetramethylbenzidine) is added in a citrate buffer pH 6.0.

The use of solid-phase techniques has led to simple and robust assays that are suitable for use throughout the world both by trained (clinical laboratories) and untrained users (e.g. home tests).

The first stage in the production of a solid-phase bound immunoreagent is the *selection* of the solid phase. *Table 1* illustrates many of the solid phases that have been used for immunoassay development.

The range of diameters for solid supports extends from microscopic particles (in the order of 100 nm) to beads (in the order of 1–10 mm) and the inner surfaces of test tubes. The smaller the particle the higher the surface area for antibody or antigen binding. The amount of particle suspension used in an assay can be easily increased if necessary for a given volume of analyte. Tubes, wells and beads have smaller surface areas exposed to similar assay volumes and so less antibody can be bound for a given assay volume. Another contrast between particles and larger solid supports is that rate of reaction with analyte is faster for smaller particles than for larger particles and surfaces (4). In spite of the advantages of particles highly successful immunoassays have been developed using all types of solid phase.

Whatever solid phase is selected their use in immunoassay requires mixing to ensure optimum performance and the fastest rate of reaction between immunoreagents. Agitation or rotation of the reaction vessel is widely used. For some magnetizable particle assays, an oscillating magnetic field around the assay cuvette has been adopted.

The second stage is the *immobilization* of an immunoreagent on the surface of a solid support by either passive adsorption or covalent coupling. Passive adsorption of antibodies and other proteins to organic polymer surfaces and glass is a well established technique which simply involves incubating the antibody in a neutral or basic buffer with the solid support (6) (Section 4). In addition a wide variety of methods have been employed for the formation of covalent bonds between solid phases and immunoreagents (7–9,16) (Section 5). This wide variety reflects the many functional groups available for coupling on both solid supports and the molecules to be coupled.

For assay development it is usually necessary to titrate the amount of immunoreagents bound to a solid phase. In the protocols below, guidance is often given as to how much antibody to add to a given solid phase. However, this should be seen as a starting point for performing a range of concentration experiments to optimize binding.

Having immobilized the immunoreagents the third stage is treatment of the solid surface to prevent or reduce the non-specific binding of labelled immunoreagents to unoccupied or unreacted sites on the surface. The processes employed are commonly referred to as *blocking* procedures (Section 7).

Table 1. Solid phase supports

Solid phase	**Form**	**Separation method**	**Immunoassay examples (refs)**
Polystyrene	Beads, particles, tubes, microtitre wells	Aspiration or decanting and washing	10–13
Derivatized polystyrene for covalent coupling	Beads, particles, microtitre wells	Aspiration or decanting and washing	14–17
Polythene	Tubes	Aspiration or decanting and washing	18
Polypropylene	Tubes, discs	Aspiration or decanting and washing	6,19–21
Polymethacrylate	Beads	Aspiration and washing	22
Nylon and nylon derivatized for covalent coupling	Membranes in immunofiltration devices	Washing through the membrane under vacuum	23
Nylon and nylon derivatized for covalent coupling	Microfine suspension	Centrifugation and washing or non-separation (see Chapter 6)	24
Copolymers	e.g. styrene and methacrylic acid copolymer particles	Centrifugation and washing	17
Cellulose	Particles	Centrifugation and washing	25–27
Agarose	Particles	Centrifugation and washing	25,27–30
Sephadex	Particles	Centrifugation and washing	25,31,32
Iron oxide	Magnetizable particles	Sedimentation on a magnetic block and washing	27,33–36
Iron oxide combined with cellulose, agarose, polystyrene or polymethacrylate	Magnetizable particles	Sedimentation on a magnetic block and washing	33,36–42
Nickel combined with cellulose	Magnetizable particles	Sedimentation on a magnetic block and washing	33,36
Chromium dioxide	Magnetizable particles	Sedimentation on a magnetic block and washing	43
Barium ferrite	Magnetizable particles	Sedimentation on a magnetic block and washing	44
Glass	Controlled pore glass particles	Centrifugation and washing	45
Microencapsulated antisera	Antibodies surrounded by semipermeable nylon, polyurea or cellulose nitrate microcapsules	Centrifugation and washing	46
Entrapped antisera	Monomers are polymerized with antibodies in the reaction mixture to form particles	Centrifugation and washing	47
Polymerized antisera	Antibodies cross-linked to form a crude precipitate	Centrifugation and washing	48

The final stage is the *storage* of the immobilized immunoreagent. It has often been found that antibodies are more stable when immobilized than in solution (49,50) (Section 8). This chapter will give information and methods for the production of immobilized immunoreagents for use in immunoassays.

2. Solid phase supports

2.1 Polystyrene

Polystyrene is an organic polymer with a hydrophobic surface. It has found widespread use as a solid support for immunoassays. Polystyrene balls of various sizes suitable to rest in the bottom of a test tube are commonly used (11). The balls are non-porous solid spheres and may have an etched surface to increase binding capacity. They are manufactured by injection moulding virgin, non-cross-linked polystyrene, and are then ground smooth. Etched surfaces are made by aqueous grinding with pumice. Polystyrene particles (also known as latex particles) with average diameters from about 40 nm to fractions of a millimetre have found widespread application in immunoassays (13,16,17) (see Chapter 5). Particles have also been made from polymers of styrene mixed with other monomers (16,17). Polystyrene microtitre plates are available with a very smooth surface at the bottom of the wells to give high optical clarity. Polystyrene test tubes are available with various coating potencies from commercial suppliers.

Proteins may be immobilized either by passive adsorption (10–13,17) (Section 4) or by covalent coupling (Section 5). Modified polystyrene surfaces have been prepared to enable the covalent coupling of molecules to the surface (14–16) (Section 5.1).

Immunoreagents immobilized on polystyrene beads, wells or tubes are used in immunoassays where the bound and the free fractions are separated by decanting or aspiration and washing. The use of polystyrene has been employed in both manual and fully automated immunoassay systems.

2.2 Other organic polymers

Test tubes, beads, particles and discs have been made of polymethacrylate (22), polyethene (18), polypropylene (6,19–21), nylon (24) and various copolymers (16,17). Proteins have been immobilized by passive adsorption to unmodified surfaces. In addition, immunoreagents have been immobilized by covalent coupling to chemically derivatized surfaces.

2.3 Cellulose particles

Cellulose is a stable linear polymer of 1,4-β-D-glucose. Due to its stability, a wide range of solvents and activation chemistries can be accomplished utilizing the abundant hydroxyl groups. Reactions may be performed in aqueous solutions (pH 3–10) and organic solvents including dimethylformamide (DMF), dimethylsulphoxide (DMSO), acetone and dioxane (51). For some applications beaded, regenerated celluloses (porous) are not suitable as the pores trap molecules preventing efficient washing and so lead to an increase in NSB.

Covalent coupling of protein immunoreagents to cellulose has been successfully employed for immunoassay development, with the favoured method being activation by 1,1′-carbonyl diimidazole (26). The addition of protein results in stable urethane bond formation (Section 5.2.2).

Suspensions of cellulose (in common with other particle suspensions) have the advantage of a large surface area for immunoreagent or antigen binding. Efficient mixing during incubation with analyte enables rapid and sensitive assays to be performed.

Cellulose particles are used in immunoassays where the bound and free fractions are separated by centrifugation and washing (25–27). Cellulose combined with magnetic particles has also been employed in immunoassays (33,36,37). Separation is achieved by sedimentation of particles on a magnetic block (Section 2.5).

2.4 Agarose, sephacryl and sephadex particles

Agarose, sephacryl and sephadex particles have been used in immunoassays in much the same way as cellulose (27–32). Agarose is a polysaccaride with alternating D-galactose and 3-anhydrogalactose. It is an uncharged hydrophilic matrix with primary and secondary alcohols for activation for covalent coupling reactions. Agarose is available in different forms (sepharoses) according to the degree of cross-linking. Sephacryl is the product of polymerizing allyl dextran with the cross-linking monomer *N,N*'-methylenebisacrylamide. The resulting structure is believed to be made up of polymeric glucose chains held together by bisacrylamide cross-links. Sephadex is a beaded dextran gel. Agarose, sephacryl and sephadex have been activated for the covalent coupling of immunoreagents by several methods (Section 5.2.2).

2.5 Magnetizable particles

Particles that are drawn to a magnetic field without being intrinsically magnetic are termed magnetizable. Suspensions of magnetizable particles have the advantage of a large surface area for immunoreagent binding without the need for centrifugation (as for other suspensions). A typical manual assay separation/wash step involves a 2 min sedimentation on a magnetic block compared with a 5 min centrifugation for cellulose particles (33).

A range of magnetizable particles have been used in immunoassays including iron oxide (33–42), nickel (36), chromium dioxide (43) and barium ferrite

(44). Iron oxide (Fe_3O_4) itself (33–35) or, more commonly, incorporated into polymers (e.g. cellulose (36,37), agarose (36,38,39), polystyrene (40–42)) has found widespread application. The size of the particles that have been employed ranges from microscopic particles (100–20000 nm) to beads (1.5 mm). For suppliers of a variety of particles see Appendix A.

Magnetizable particles are employed in both manual and fully automated assays (e.g. Serono MAIAclone™, Chiron/Ciba Corning ACS 180™, Boehringer Elecsys™).

2.6 Membranes

Immunoreagents have been immobilized on nylon membranes or on polymer particles dried onto membrane surfaces (23). The bound fraction of an assay is held on the membrane whilst unbound components are washed through (e.g. Pierce Easy Titre™ Enzyme linked immunofilter assay—ELIFA, Hybritech ICON™, Kodak SureCell™).

Another application of membranes is in dry surface immunoassays. Immunoreagents and other assay components are immobilized or impregnated into permeable or porous membranes (52). The membranes are dried and incorporated into single layer or multilayer devices. Such membranes include cellulose, gelatin, agarose and methylcellulose. On addition of a liquid sample the analyte penetrates the layers and binds to the specific antibodies present. Detection by automated systems usually depends on either reflectance photometry or fluorimetry. Protocols detailing the preparation of these devices are not given in this book.

Semipermeable membranes have been used for the microencapsulation of antisera. The antibodies are not attached to the membrane but are held in solution within a capsule. The membrane (cut off limit about 20 kDalton) prevents the antibodies escaping whilst small molecules such as drugs and low molecular weight peptides may enter the capsule from the sample. Separation is by centrifugation and washing (46).

3. Preparation of solid supports

It is possible to synthesize many types of solid phase in a chemistry laboratory. However, it is much more convenient and often less expensive to purchase them from commercial suppliers (see Appendices A and B).

Methods are given for the preparation of iron oxide magnetizable particles (*Protocols 1–2*) and iron oxide combined with polystyrene and polymethacrylate (*Protocols 3–4*). The inclusion of methacrylic acid in the two polymerization mixtures gives carboxylic acid functional groups on the particle surfaces (34,35).

Protocol 1. Preparation of iron oxide particles (NaOH precipitation)

1. Mix 125 ml 0.4 mol/litre $FeCl_2$ and 0.25 mol/litre $FeCl_3$ with 125 ml of 5 mol/litre NaOH solution. Stir the mixture for 2 min at 50 °C.
2. A black, magnetizable iron oxide suspension forms. Decant the precipitate and wash with water until a pH of 7–8 is reached.

Protocol 2. Preparation of iron oxide particles (NH_4OH precipitation)

1. Cool 150 ml 0.2 mol/litre $FeCl_2$ and 150 ml 0.35 mol/litre $FeCl_3$ to 10 °C and then mix.
2. Add 100 ml 5 mol/litre NH_4OH solution precooled to 10 °C at a rate of 5 ml/s with continuous stirring. Stir for 1 h in an ice bath while a precipitate forms.
3. A black, magnetizable iron oxide suspension is obtained. Wash with 0.9% NaCl until the supernatant is neutral and then twice with water.

Protocol 3. Preparation of polyacrylamide magnetizable particles

Perform all reactions under a nitrogen atmosphere.

1. Place 5 g magnetizable iron oxide particles in 400 ml water in a round bottom flask equipped with a stirrer (agitation speed 700 rpm) and a condenser.
2. Add with stirring 0.5 g of sodium dodecyl sulphate (SDS) as an emulsifying agent, 3 g of methacrylic acid, 9 g of 2-hydroxyethylmethacrylate, 3 g of acrylamide, 1.5 g of *N,N'*-methylene-bisacrylamide. Heat the reaction mixture to 70–80 °C.
3. Add 0.1 g of potassium persulphate ($K_2S_2O_8$) as the free radical initiator and stir for 7 h (the iron oxide particles are coated with copolymers).
4. Add a further 3 g of methacrylic acid and 0.05 g $K_2S_2O_8$ and stir for 5 h.
5. Wash the resulting polyacrylamide magnetizable particles exhaustively with water.

Protocol 4. Preparation of polystyrene magnetizable particles

Perform all reactions under a nitrogen atmosphere.

1. Place 5 g magnetizable iron oxide particles in 400 ml water in a round bottom flask equipped with a stirrer (agitation speed 700 rpm) and a condenser.
2. Add with stirring 1 g of sodium dodecyl sulphate (SDS) as an emulsifying agent, 3 g of methacrylic acid, 6 g of 2-hydroxyethyl-methacrylate, and 0.1 g $K_2S_2O_8$. Heat the reaction mixture to 70–80 °C. Add 7.5 g of styrene and 1 g of divinylbenzene with stirring for 8–10 h (the iron oxide particles are coated with copolymers).
3. Add a further 3 g of methacrylic acid and 0.05 g of $K_2S_2O_2$. Stir for 5 h.
4. Wash the resulting polystyrene magnetizable particles three times with methanol or ethanol and then exhaustively with water.

4. Passive adsorption

One of the most frequently used methods of immobilizing protein immunoreagents is by passive adsorption to a solid surface (6) (*Figure 1*). The binding is thought to be mainly due to hydrophobic forces. This approach is suitable for the binding of most proteins (except small peptides) to organic polymer surfaces (e.g. antibody to polystyrene) and glass. Small molecules may be immobilized by passive adsorption provided they have been covalently coupled to a protein (e.g. digoxin covalently bound to bovine serum albumin, BSA). It is not usually recommended that the protein be the same as that used to make the immunogen for antibody production. This is because some of the antibodies produced will be directed against the immunogen protein as well as the hapten.

Ideally, optimization of antibody concentration, coating buffer concentration (0.1–0.01 mol/litre), pH and temperature should be carried out for each immobilized immunoreagent preparation. A widely adopted procedure

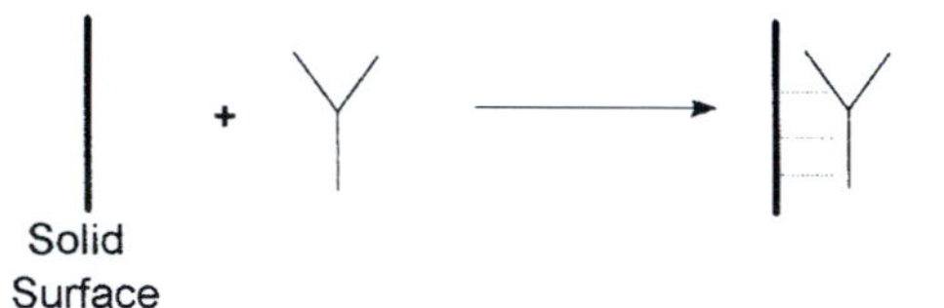

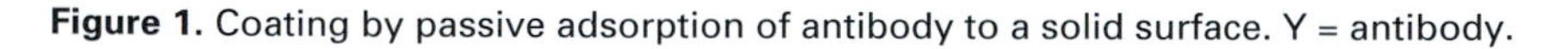
Figure 1. Coating by passive adsorption of antibody to a solid surface. Y = antibody.

dilutes the antibody in 50 mmol/litre sodium carbonate pH 9.2–9.6 and this serves as a good starting point for assay development (6). Satisfactory results have also been achieved with phosphate buffered saline pH 7.4 and Tris buffered saline pH 8.5.

Direct adsorption of antibody to polystyrene wells, beads and tubes will often be adequate for assay development. However, in some instances (which can only be determined empirically) optimum assay performance, notably precision, is obtained by indirect binding of the primary antibody (Section 6.1).

Methods for the passive adsorption of antibody to polystyrene microtitre wells, beads and tubes are given below (*Protocols 5–7*). These protocols are suitable for use with other organic polymer surfaces apart from polystyrene (e.g. polypropylene tubes etc.).

Protocol 5. Passive coating of polystyrene microtitre wells with antibody

Materials

1. Coating buffer: 0.05 mol/litre Na_2C0_3 pH 9.6 with 0.1% sodium azide.
2. Wells: Nunc Maxisorp microtitre plates.
3. Wash buffer: Phosphate buffered saline, PBS, (0.025 mol/litre phosphate, 0.15 mol/litre NaCl, pH 7.4) with 0.01% triton X-100.
4. Blocking solution: 5% solution (w/v) of non-specific protein such as hydrolysed gelatin or bovine serum albumin (BSA) in coating buffer.
5. Glazing solution: 2% mannitol in distilled water.

Method

1. Dilute antibody in coating buffer and add 200 μl/well.
2. Leave at 25 °C over 3 days (or 24 h 37 °C) in a humid atmosphere (e.g. closed plastic box with wet tissue paper).
3. Add 50 μl blocking solution and leave for 24 hours in a humid atmosphere.
4. Aspirate and wash the wells 4 times with wash buffer.
5. Add 300 μl/well glazing solution.
6. Aspirate and vacuum dry the wells.
7. Store the wells with a desiccant in air tight bags or containers at 4 °C.

Protocol 6. Passive coating of polystyrene beads with antibody

Materials

1. Coating buffer: 0.05 mol/litre Na_2CO_3 pH 9.6 with 0.1% sodium azide.
2. Beads: Polystyrene balls 6.4 mm diameter, specular finish No.P201 from NBL Gene Sciences.
3. Wash buffer: Phosphate buffered saline, PBS, (0.025 mol/litre phosphate, 0.15 mol/litre NaCl) with 0.01% triton X-100.
4. Blocking solution: 5% solution (w/v) of non-specific protein such as hydrolysed gelatin or bovine serum albumin (BSA) in coating buffer.
5. Glazing solution: 2% mannitol in distilled water.

Method

1. Dilute antibody in coating buffer.
2. Add antibody solution to cover the beads, typically 125 ml solution to cover 1000 beads in a 250 ml container (screw top plastic or container that can be sealed).
3. Before closing container degas the beads under vacuum (to remove the air bubbles on the surface of the beads).
4. Close the container and store the beads at 25 °C over 3 days (or 24 h at 37 °C).
5. Aspirate the coating solution and add the blocking solution to cover the beads—typically 125 ml for 1000 beads. Leave in closed container for 24 h at 25 °C.
6. Aspirate and wash the beads 3 times with washing buffer.
7. Add glazing solution to cover the beads and aspirate.
8. Vacuum dry the wet beads.
9. Store the beads with a desiccant in air tight bags or containers at 4 °C.

Protocol 7. Passive coating of polystyrene tubes with antibody

Materials

1. Coating buffer: 0.05 mol/litre Na_2CO_3 pH 9.6 with 0.1% sodium azide.
2. Tubes: Greiner medium binding tubes, 12.0/75 mm, 115001.
3. Wash buffer: Phosphate buffered saline, PBS, (0.025 mol/litre phosphate, 0.15 mol/litre NaCl) with 0.01% triton X-100.

4. Blocking solution: 5% solution (w/v) of non-specific protein such as hydrolysed gelatin or bovine serum albumin (BSA) in coating buffer.
5. Glazing solution: 2% mannitol in distilled water.

Method

1. Dilute antibody in coating buffer.
2. Add antibody solution to the tubes (300 μl/tube). Leave at 25 °C over 3 days (or 24 h at 37 °C) in a humid atmosphere (e.g. closed plastic box with wet tissue paper).
3. Add 50 μl blocking solution and leave for 2–3 h at 25 °C in a humid atmosphere.
4. Decant and wash the tubes (3 × 2 ml/tube) with wash buffer.
5. Add 300 μl/tube glazing solution.
6. Decant and vacuum dry the tubes.
7. Store the tubes with a desiccant in air tight bags or containers at 4 °C.

5. Covalent coupling

A large variety of chemical methods have been employed for the covalent coupling of molecules to solid surfaces (7–9,16,53). The aim of all the methods is to form a stable covalent link between the immunoreagent and the solid surface without damaging immunoreactivity. As with passive adsorption it is necessary to optimize the amount of immunoreagent bound on a surface.

The chemistry involved depends on which solid support has been chosen and the properties of the molecule to be coupled. Molecules may be coupled to solid phases with suitable groups for covalent bond formation. Such groups include: amines, carboxylic acids, active esters, aldehydes, carbohydrate hydroxyls, epoxides, sulphonyl chlorides and maleimides. *Table 2* summarizes many of the functional groups that will form covalent bonds between solid surfaces and immunoreagents. For more detailed discussion of the reactions see the Sections 5.2.1–5.2.5. Some solid phases are without these groups and so first require derivatization to introduce them (Section 5.1).

5.1 Derivatization of solid supports

A number of commonly used solid phases (e.g. iron oxide, glass, polystyrene and other polymers) do not have functional groups on their surfaces suitable for covalent coupling reactions with immunoreagents. Therefore it is necessary to derivatize the surfaces with the aim of introducing suitable groups on the surface.

Polystyrene beads have been derivatized for covalent coupling reactions by brief treatment with chlorosulphonic acid. This introduces a sulphonyl

Table 2. Functional groups on solid surfaces and immunoreagents

Solid phase functional groups	Immunoreagent functional groups that will couple	Comments
Amines (5.2.1)	Active esters, isocyanates, isothiocyanates, sulphonyl halides, anhydrides, mixed anhydrides, acid halides, epoxides, aldehydes	Apart from active esters these groups are not used for coupling proteins to solid phase amines. However, they are commonly employed for small molecules.
Hydrazides (5.2.4)	Aldehydes	In addition to hapten aldehydes glutaraldehyde is used to introduce aldehyde groups onto hydrazide-functionalized surfaces (see aldehydes below).
Carbohydrate hydroxyls (5.2.2)	Amines	Only used if the hydroxyls are activated to nucleophilic attack (see CDI-activated, tosylated or tresylated hydroxyls).
Carbonyl diimidazole (CDI) activated hydroxyls (5.2.2)	Amines	Preferred method for coupling to cellulose. Forms a stable urethane bond.
Tosylated or tresylated hydroxyls (5.2.2)	Amines	Forms a stable substituted amine bond.
Carboxylic acids (5.2.3)	Amines	Only used if the carboxylic acid is activated to nucleophilic attack, see active esters below.
Active esters (5.2.3)	Amines	Frequently formed by EDC or EDC with NHS. Forms stable amide bonds.
Aldehydes (5.2.4)	Amines	Glutaraldehyde is used to modify amine and hydrazide functionalized solid phases. Addition of amines followed by reduction results in stable substituted amine bonds.
Epoxides	Amines	Forms stable substituted amine bonds.
Sulphonyl halides	Amines	Forms stable sulphonamide bonds.
Maleimides (5.2.5)	Thiols	Forms stable thioether bonds.

EDC = 1-ethyl-3-(3-dimethylaminopropyl)carbodiimide; Tosyl = *p*-toluenesulphonyl chloride; Tresyl = 2,2,2-trifluoroethane sulphonyl chloride; NHS = *N*-hydroxysuccinimide.

chloride group to the phenyl ring which reacts with amines or hydrazine. Alternatively, brief treatment of polystyrene beads with a mixture of nitric and sulphuric acids followed by reduction introduces an amine group to the phenyl ring which reacts with active esters, isothiocyanates, isocyanates etc. Amino- and hydrazide-functionalized 1/4″ beads are commercially available (Pierce).

Microtitre wells are unable to withstand any of the chemistry described for beads without losing their optical properties. There are two methods of introducing functional groups onto the plate surface: passive adsorption of a polymer with the desired functional groups and chemically modified plates supplied commercially: Nunc, secondary amine, and Costar, primary amine and carboxylic acid.

Another approach to derivatizing polymers has been adopted for polystyrene and polymethacrylate. Methacrylic acid may be added to the polymerization mixture. This gives carboxylic acid functional groups on the polymer surface (34,35) (see *Protocols 3* and *4*).

Iron oxides and glass may be derivatized by amino silyl compounds to incorporate amino groups on the surface (53, 54) (*Figure 2*, *Protocols 8* and *9*).

Alternatively, iron oxide particles have been combined with carbohydrate polymers (e.g. cellulose and agarose). The resultant particles have carbohydrate hydroxyl groups for activation (Section 5.2.2) combined with the magnetizable properties of iron oxide (33,36,38,39).

Compounds used for derivatization, apart from enabling covalent coupling, also introduce varying spacer groups between the solid support and the immunoreagent (see Section 5.2 for further discussion).

Protocol 8. Preparation of silanized magnetizable particles by an acidic organic silanization

Perform all reactions under a nitrogen atmosphere with stirring.

1. Wash 5 g of iron oxide particles three times with methanol or ethanol to remove most of the water in the particle suspension.
2. Suspend the particles in 150 ml of methanol or ethanol containing approximately 0.5% water. Add 10 g of 3-(2-aminoethylamino)propyltrimethoxysilane (Fluka) and 2 g of orthophosphoric acid.
3. Stir the mixture at about 2000 rpm for 15 min and at about 1000 rpm for 2 h at room temperature.
4. Mix the contents with 200 ml of glycerol in a round bottom flask equipped with a stirrer and a condenser, and heat to 180 °C for 1 h and then cool to room temperature.
5. Wash the glycerol particle slurry exhaustively with water.

Protocol 9. Preparation of silanized magnetizable particles by an acidic aqueous silanization

Perform all reactions under a nitrogen atmosphere with stirring.

1. Mix 5 g of iron oxide particles with 400 ml of 10% solution of 3-(2-aminoethylamino)propyltrimethoxysilane (Fluka) in a round bottom flask equipped with a stirrer.
2. Adjust the pH to about 5 with glacial acetic acid. Heat the mixture to 90–95 °C for 2 h with stirring (about 2000 rpm for 15 min and at about 1000 rpm for the rest of the time).
3. Mix the contents with 200 ml of glycerol in a round bottom flask equipped with a stirrer and a condenser and heat to 180 °C for 1 h and then cool to room temperature.
4. Wash the silanized magnetizable particles three times with water, three times with methanol and three times with water.

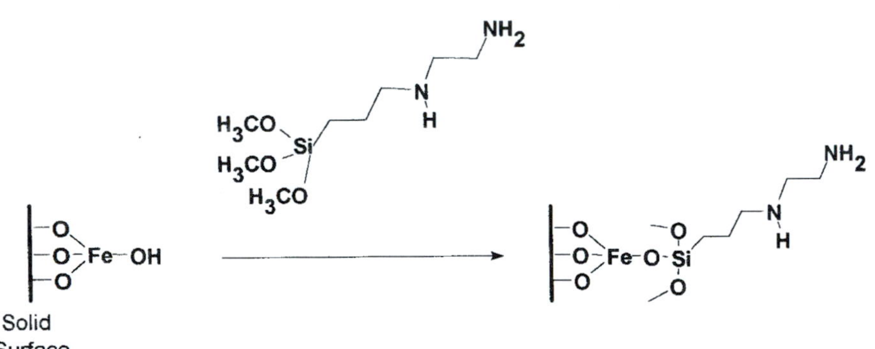

Figure 2. The derivatization of iron oxide with 3-(2-aminoethylamino)propyltrimethoxysilane to incorporate amino groups on the surface.

5.2 Covalent coupling reactions with solid supports

This section describes the covalent coupling of molecules to solid surfaces that either have suitable groups for covalent bond formation or have been derivatized to include such groups (7–9,16,53).

The process of derivatization and coupling of immunoreagents to solid phases will lead to varying lengths of spacer arms (upwards from zero atoms) between the surface and the coupled molecule. It is recommended that the spacer arm is of a different chemical structure than that used in the production of antisera as otherwise this can lead to raised NSB. Also, for some assays, long hydrophobic spacer arms may increase NSB by forming hydrophobic bonds with proteins.

For the coupling of antibodies and many proteins it is not often necessary to employ spacer arms. However, for small molecules and low molecular weight proteins, spacer arms are usually employed (55,56). This is because an antibody can be prevented sterically from binding a small molecule close to the solid surface.

The variety of chemistries available enables molecules to be coupled with different orientations. For example, a low molecular weight peptide with one cysteine residue and one lysine could be coupled via the thiol group (e.g. to a maleimide-functionalized solid phase, Section 5.2.5) or the amino group (e.g. to an active ester-functionalized solid phase, Section 5.2.3). If the cysteine and lysine are found at different ends of the peptide, different immunoreactivities with antibodies will result.

5.2.1 Amine-functionalized solid surfaces

Amine-functionalized surfaces enable a variety of covalent coupling reactions to be performed. The reactions may be divided into two strategies. The first strategy involves forming reactive groups susceptible to nucleophilic substitution on the molecule to be attached to the solid phase. A frequently used approach for the coupling of proteins (e.g. antibodies) is the formation of active esters by the addition of 1-ethyl-3-(3-dimethylaminopropyl)carbodiimide (EDC) (57–59). Active esters on the protein react to form stable amide bonds with solid phase amines (*Figure 3*). Although the EDC method has been employed successfully many times for the coupling of antisera to solid phases (*Protocols 10* and *11*) it can be accompanied by some loss of immunoreactivity due to cross-linking of the antisera (i.e. amines and carboxylic acids on different protein molecules couple together). Where necessary, cross-linking can be avoided by prior activation of a solid phase with carboxylic acid groups by EDC with or without *N*-hydroxysuccinimide. Following activation the protein immunoreagent may be coupled (Section 5.2.3).

Protocol 10. Carbodiimide method for coupling antiserum (unpurified) to amino-functionalized magnetizable particles

1. Wash 0.2 g of 3-(2-aminoethylamino)propyltrimethoxysilane derivatized magnetic particles (*Protocol 8* or *9*) three times with water (30 ml/wash) and resuspend in 8 ml of water.
2. Add 0.2 ml antiserum. After mixing for 2 min add 40 mg 1-ethyl-3-(3-dimethylaminopropyl)carbodiimide (EDC). Adjust the pH to 5.6 with 0.1 mol/litre HCl or 0.1 mol/litre NaOH.
3. Rotate the mixture for about 24 h at room temperature.
4. Wash the magnetizable immobilized antibody three times with phos-

Protocol 10. *Continued*

phate buffered saline (PBS, 0.025 mol/litre phosphate, 0.15 mol/litre NaCl, pH 7.4) containing 0.1% bovine serum albumin (BSA) and three times with water, then three times with PBS and once with Tris buffer (0.05 mol/litre, pH 7.5) containing 0.004 mol/litre EDTA (30 ml/wash).

5. Resuspend in PBS containing 1% BSA, 0.1% sodium azide and store at 4°C.

Protocol 11. Carbodiimide method for coupling antibody (partially purified) to amino-functionalized magnetizable particles

1. Wash 0.2 g of 3-(2-aminoethylamino)propyltrimethoxysilane derivatized magnetic particles (*Protocol 8* or *9*) three times with phosphate buffered saline (PBS, 0.025 mol/litre phosphate, 0.15 mol/litre NaCl, pH 7.4) containing 0.1% BSA and 0.1% Tween 20, then wash once with 0.05 mol/litre 2-(*N*-morpholino)ethane sulphonic acid pH 6.0 (MES). Use 30 ml of solutions for each wash.
2. Resuspend the particles in 15 ml MES. Add 2 ml purified antibody (partially purified by caprylic acid precipitation).
3. Rotate for 30 min at room temperature and add 20 mg EDC. Rotate for a further 24 h.
4. Wash the magnetizable immobilized antibody five times with PBS containing 0.1% bovine serum albumin (BSA) and 1% Tween 20, then five times with PBS containing 0.1% BSA and 0.1% Tween 20 (30 ml/wash).
5. Resuspend in PBS containing 1% BSA, 0.1% sodium azide and store at 4°C.

The favoured approach for covalently coupling many small molecule derivatives is to activate them to nucleophilic substitution by amines on the solid surface. The active groups on small molecules that may be coupled to amines include active esters, isocyanates, isothiocyanates, sulphonyl chlorides, mixed anhydrides, acid chlorides and epoxides (60). For the numerous carboxylic acid derivatives formation of active esters by EDC with *N*-hydroxysuccinimide (NHS) has found frequent use (58,60,61) (*Figure 4*). A method is given for the covalent coupling of an *N*-hydroxysuccinimide ester to an amino-functionalized solid phase (microtitre wells) (*Protocol 12*). For situations where a water soluble hapten derivative is necessary, active esters may be formed from EDC with *N*-hydroxysulphosuccinimide (sulphoNHS)

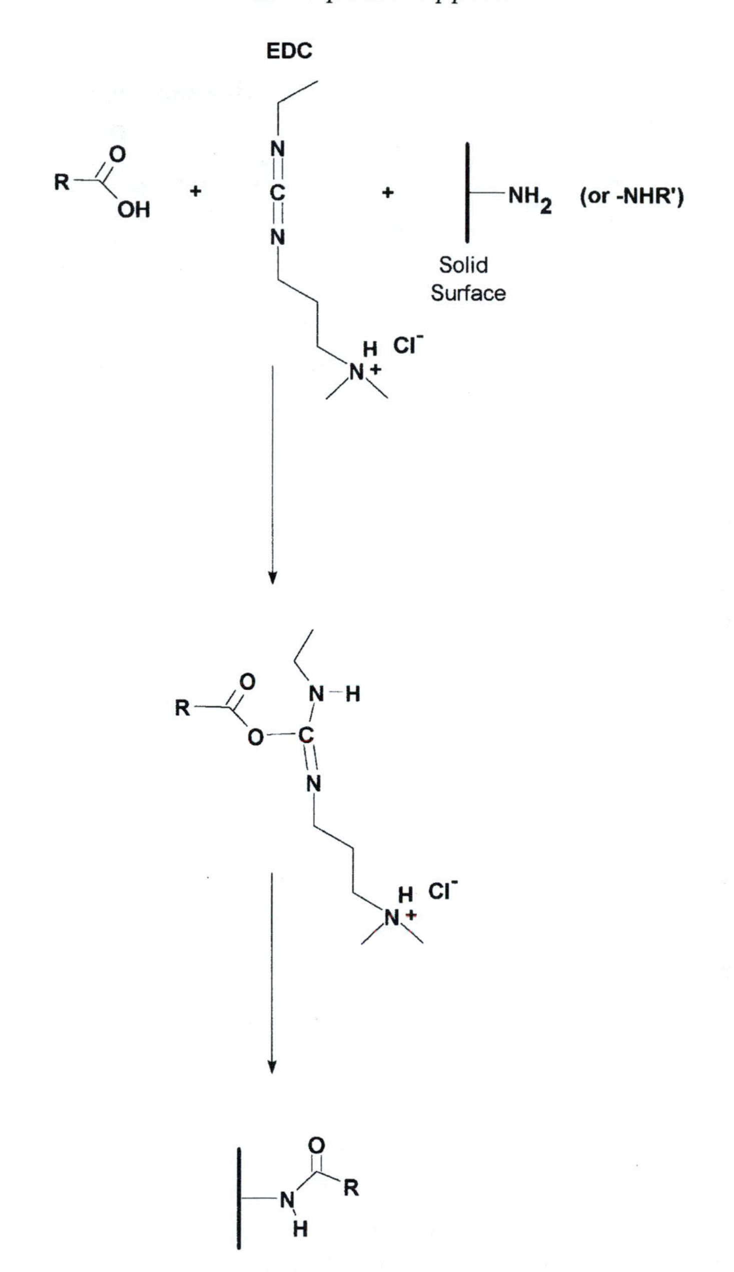

Figure 3. The activation and coupling of proteins to amino-functionalized surfaces by the carbodiimide method. EDC = 1-ethyl-3-(3-dimethylaminopropyl)carbodiimide, RCO_2H = protein with carboxylic acid groups.

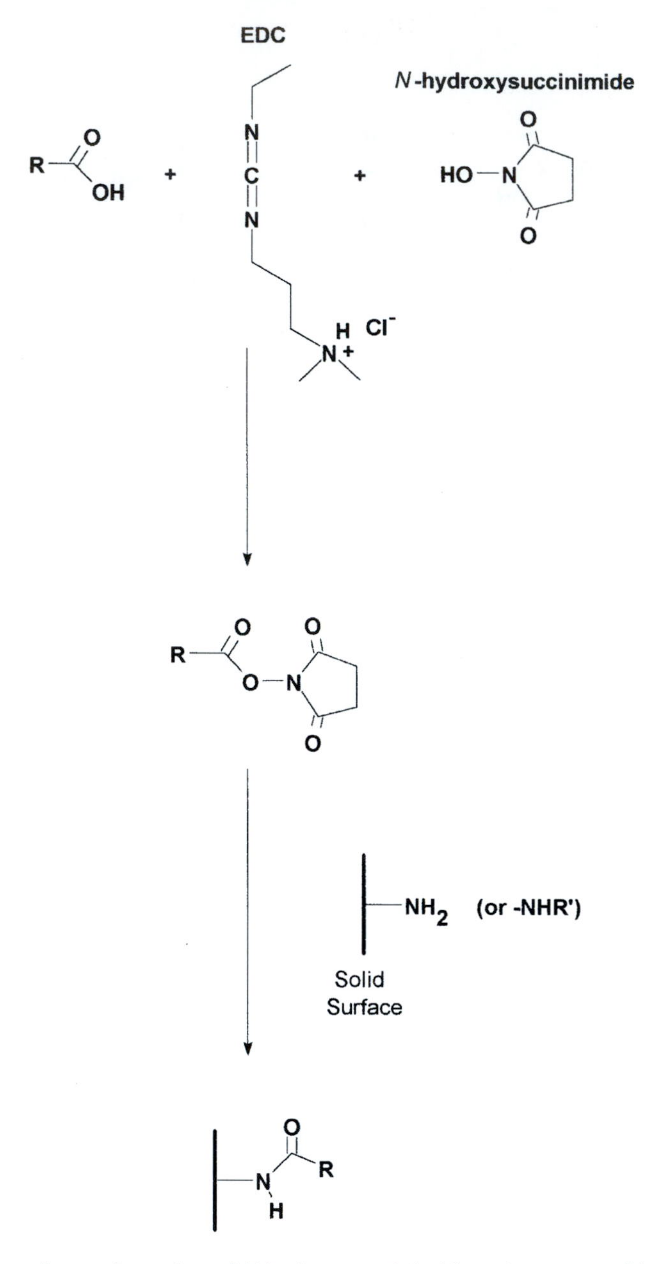

Figure 4. The formation and reaction of *N*-hydroxysuccinimide active esters with amino-functionalized surfaces. EDC = 1-ethyl-3-(3-dimethylaminopropyl)carbodiimide, RCO_2H = small molecule with a carboxylic acid group.

(62,63) instead of NHS/EDC. The sulphonic acid group on sulphoNHS (Pierce) confers additional water solubility on the activated molecule.

Protocol 12. Covalent coupling of *N*-hydroxysuccinimidobiotin to secondary amine functionalized microtitre wells

Introduction

Secondary amine functional groups on 2 nm long spacer arms on the surface of microtitre wells at an approximate concentration of 10^{14} cm^{-2} are reacted with a biotin *N*-hydroxysuccinimide active ester. Any other active ester may be substituted.

Materials and solutions

1. Nunc covalink microtitre plates.
2. *N*-Hydroxysuccinimidobiotin (Sigma or Pierce).
3. Phosphate buffered saline (PBS, 0.1 mol/litre phosphate, 0.15 mol/litre NaCl) pH 7.4.
4. Blocking buffer: 0.5% bovine serum albumin (BSA) solution in PBS.
5. Washing buffer: PBS with 0.05% triton X-100.

Method

1. Prepare a 10 mg/ml *N*-hydroxysuccinimidobiotin solution in dry dimethylsulphoxide (DMSO) freshly before use.
2. Dilute the DMSO solution of *N*-hydroxysuccinimidobiotin to 250 μg/ml in PBS.
3. To as many wells as desired add 100 μl of the *N*-hydroxysuccinimidobiotin solutions. Cover the wells and incubate at room temperature overnight.
4. Aspirate the wells and add 300 μl blocking solution. Cover the wells and incubate for 1 h.
5. Aspirate and wash each well with washing buffer four times.

The levels of coupling can be monitored by adding 100 μl avidin–HRP solution and detecting the binding with an HRP substrate.

The second strategy involves modifying the solid phase amine groups to present alternative functional groups (e.g. carboxylic acid, aldehyde, maleimide) on the surface (*Figure 5*).

Aldehyde groups can be introduced by the addition of glutaraldehyde (a 'homobifunctional' reagent) to solid surface amines. The aldehyde group from one end of glutaraldehyde forms a Schiff base with the solid phase

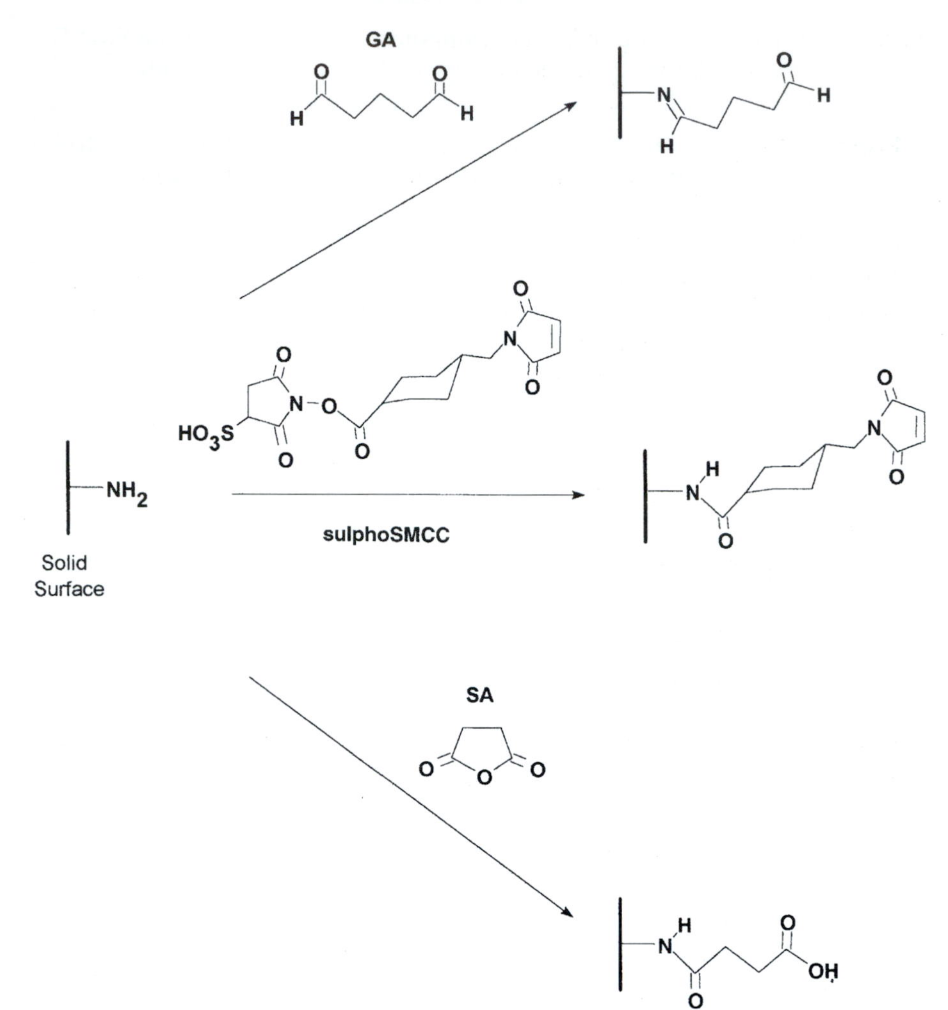

Figure 5. The introduction of alternative functional groups to an amino-derivatized solid surface using glutaraldehyde (GA), sulphosuccinimidyl-4-(*N*-maleimidomethyl)cyclohexane-1-carboxylate (sulphoSMCC) and succinic anhydride (SA).

amine. This is subsequently reacted with the molecule to be coupled to the solid phase (which should also contain amine groups) (see Section 5.2.4). For the best results aldehyde-modified surfaces should not be stored but used immediately to couple the molecule to be bound (64,65). Methods for this approach are described in *Protocols 23* and *24*.

Carboxylic acid groups are introduced using cyclic anhydrides (66). The surface amine group forms a covalent bond with one of the carbonyl groups and so opens the ring. The resultant carboxylic acid is then activated for coupling as described later (Section 5.2.3). Carboxylic acid modified surfaces may be stored (sealed with desiccant at 4°C) before coupling of the molecule to be bound is performed. Methods for this approach are described in *Protocols 13* and *14.*

Maleimide groups are introduced using sulphosuccinimidyl-4-(*N*-maleimidomethyl)cyclohexane-1-carboxylate (sulphoSMCC, a 'heterobifunctional' reagent) (67,68). The solid phase amine groups react with the active ester groups (sulphoNHS ester, see earlier in this section) to form stable amide bonds. The maleimide groups on the surface are then coupled to sulphydryl groups on molecules (Section 5.2.5). In our experience the reaction with the thiol should be performed immediately following the coupling to the surface. A method for this approach is described in *Protocol 25.*

Methods for the introduction of carboxylic acids using glutaric anhydride and succinic anhydride are given below (*Protocols 13* and *14*). Glutaric anhydride reacts in the same way as succinic anhydride but has one more CH_2 group in the ring. Protocols for the introduction of aldehydes and maleimides are given in later sections (Section 5.2.4, *Protocols 23* and *24*; Section 5.2.5, *Protocol 25*).

Protocol 13. Glutaric anhydride method for introducing carboxylic acid functional groups onto amino-functionalized magnetic particles

1. Wash 2 g of 3-(2-aminoethylamino)propyltrimethoxysilane-derivatized magnetic particles (*Protocol 8* or *9*) four times with 0.1 M $NaHCO_3$.
2. Adjust the volume to about 50 ml and add 0.2 g glutaric anhydride. Mix the particles for 2 h at room temperature by gently rotating the reaction bottle.
3. Wash the particles twice with 0.1 M $NaHCO_3$ and repeat the addition of glutaric anhydride.
4. Wash the carboxylic acid derivatized magnetizable particles five times with water.

Protocol 14. Succinic anhydride method for the introduction of carboxylic acid functional groups onto an amine-derivatized polystyrene bead surface

1. Add 30 'alkylamine beads' (Pierce) to 50 ml 50 mmol/litre phosphate buffer pH 6.0.
2. Add 2.0 g succinic anhydride to the solution covering the beads. Gently rotate overnight at room temperature.
3. Wash the beads in a Buchner funnel (no filter paper) with 200 ml water.

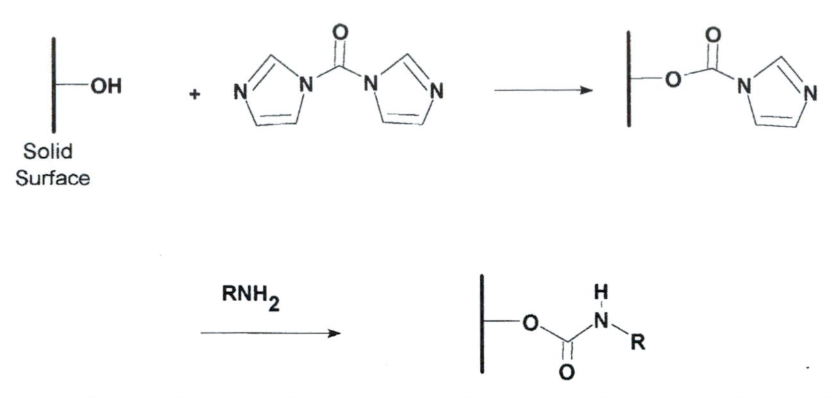

Figure 6. The coupling of amine-functionalized molecules (e.g. proteins) to hydroxyl groups (e.g. on cellulose) activated by carbonyl diimidazole (CDI). RNH_2 = amine.

5.2.2 Hydroxyl-functionalized solid surfaces

Carbohydrate hydroxyl groups (e.g. on cellulose, agarose, sephacryl and sephadex) are commonly activated by the carbonyl diimidazole (CDI) (26,40,69,70), cyanogen bromide (CNBr) (71) or *p*-toluenesulphonyl chloride (tosyl chloride) (72–74) methods. Other methods include the use of divinyl sulphone (75), 2,2,2-trifluoroethane sulphonyl chloride (tresyl) (40,72–74), and periodate (76).

CDI activation is the preferred method for cellulose (26). The hydroxyl groups are converted to imidazole carbamates which react with amines to form stable uncharged urethane bonds (*Figure 6*). CDI-activated cellulose is relatively stable once made and can be stored as a dry powder (sealed from moisture at 4°C for 3 months). However, the carbonyl diimidazole itself is very susceptible to hydrolysis and dry solvents are essential for successful activation of the matrix. Any appearance of CO_2 bubbles during activation indicates the presence of water and low yields will result (26). A wide range of reaction conditions are tolerated by cellulose for activation (including aqueous pH 3–10, DMF, DMSO, acetone and dioxane). Methods are given for the CDI activation and coupling of antiserum to cellulose particles (*Protocols 15* and *16*).

Protocol 15. Activation of cellulose by carbonyl diimidazole

Materials

1. Microparticulate cellulose, Sigmacell Type 20 (Sigma).
2. 1,1′-Carbonyl diimidazole (Sigma).
3. Acetone AnalaR (BDH).
4. Glass fibre filter, GF/A, 2 μm retention (Whatman).

This method is also applicable to magnetic cellulose/iron oxide particles.

Safety

1. Acetone is highly flammable and should be stored, handled and disposed of with no exposure to flame or spark sources. Disposal into drains can lead to explosion.
2. Wear a dust mask when handling cellulose and CDI-activated cellulose powders.

Method

1. Weigh 100 g cellulose into a 1 litre flask, add 400 ml acetone, mix and stir vigorously.
2. Weigh 12.2 g carbonyl diimidazole (CDI) and dissolve in 100 ml acetone.
3. Add the CDI solution to the cellulose and continue mixing for 1 h.
4. Collect the cellulose over a GF/A filter in a Buchner funnel.
5. Wash the cellulose in the funnel with at least 2.5 litre acetone passed through slowly.
6. Allow the activated cellulose to dry in air overnight. Store in a tightly sealed container at –20 °C.

Magnetizable particles

The same activation procedure can be applied to magnetizable cellulose/ iron oxide particles provided that the particles are transferred from aqueous suspension to acetone. This is achieved by successively washing the particles with 20, 40, 60 and 80% acetone:H_2O mixtures and then three times with anhydrous acetone. After resuspending in acetone, CDI dissolved in acetone can be added according to the above procedure.

Protocol 16. Coupling of antiserum to CDI-activated cellulose

This protocol may be scaled down for making smaller quantities of cellulose immobilized antibody. The method may be used to couple antiserum to magnetizable cellulose particles.

Safety

1. Wear a dust mask when handling dry activated cellulose.
2. Wear eye protection when handling solutions.

Materials

1. Reaction vessel: 1 litre polycarbonate centrifuge bottle with a screw top. For smaller scale preparations, screw top containers which can be centrifuged are required.

Protocol 16. *Continued*

2. Use purified, sterile filtered water throughout.
3. Prepare all solutions fresh for each preparation.
4. 0.5 mol/litre phosphate buffer (stock): Weigh 71.6g $NaH_2PO_4 \cdot 2H_20$ and 13.53 g $NaH_2PO_4 \cdot H_20$, add 2 g NaN_3 in water and make the volume to 1 litre. Check pH = 7.3–7.4 (7.4 when diluted). Store at room temperature.
5. 50 mmol/litre phosphate buffer: Dilute the 0.5 mol/litre phosphate buffer 10-fold with water, add sodium azide to a concentration of 0.1%.
6. Antibiotic phosphate buffer: 50 mmol/litre phosphate buffer containing 0.34g/litre chloramphencol and 0.1 g/litre neomycin sulphate.
7. Protein phosphate buffer: 50 mmol/litre phosphate buffer containing 1% bovine serum albumin.
8. Borate buffer 0.05 mol/litre, pH 8: Dissolve $Na_2B_4O_7$ in water to a concentration of 19.1 g/litre and adjust pH with HCl (before diluting to final volume).
9. Ethanolamine, 0.5 mol/litre, pH 9.5: Dilute ethanolamine in water to a concentration of 30 ml/litre and adjust pH with HC1 (before diluting to volume).
10. Sodium acetate 0.1 mol/litre, pH 4: Dissolve $CH_3COONa \cdot 3H_20$ in water to a concentration of 13.6 g/litre and adjust pH with concentrated HC1 (before diluting to volume).
11. Sodium bicarbonate, 0.5 mol/litre, pH 8: Dissolve $NaHCO_3$ in water to a concentration of 42 g/litre and adjust pH with HCl if necessary.
12. Diluent buffer, for solid phase. To prepare a 2 litre volume, weigh the following materials:

    14.32 g $NaH_2PO_4 \cdot 2H_2O$

    2.71 g $NaH_2PO_9 \cdot H_2O$

    25.0 g Bovine serum albumin

    4.4 g Bovine gamma globulin

    12.0 g $Na_2EDTA \cdot 2H_20$

    2.0 g Sodium azide

Dissolve with stirring in about 1600 ml water, dilute to 2 litres and add 12.5 ml Tween 20 and mix to dissolve. Finally filter the buffer through a 0.2 μm filter.

Method

Basic mix: Add antiserum diluted to 600 ml (final) with borate buffer and 150 g activated cellulose. Smaller amounts may be used with the same ratio of borate:activated cellulose.

1. Dilute the antibody preparation to the appropriate concentration with borate buffer (e.g. 100 ml unpurified antiserum up to 600 ml with borate). [For magnetizable cellulose the volume of liquid added to the dry solid should be twice that for cellulose (so as to make a mobile suspension).]
2. Weigh 150 g of CDI-activated cellulose into a disposable container.
3. Transfer the vial of antiserum into a 1 litre centrifuge bottle, rinse the vial with borate buffer and add further borate buffer to a final volume of 600 ml.
4. To the bottle containing the diluted antiserum add 150 g of the activated cellulose using a powder funnel.
5. Close the bottle and mix the contents by shaking. Place the bottle on a roller mixer and leave mixing (at room temperature) for at least 20 h (or over a weekend).
6. Sediment the solid phase by centrifuging at 2000 rpm for 20 min and decant off the supernatant.
7. Wash the solid phase antiserum:
   (a) The cellulose-linked antibody is washed several times with buffer solutions.
   (b) For each wash, add the solution to the solid phase (approx. 250–300 ml per bottle) and mix on a roller mixer for the specified times below (mix longer if resuspension should not be complete). [For small-scale preparations, typically 1 g of cellulose, then the wash volumes should be 10 ml/wash.]
   (c) Sediment the solid phase by centrifuging at 2000 rpm for 20 min and decant off the supernatant.
   (d) Wash the solid phase as follows:

      Ethanolamine, mix for 20 min (this blocks any unreacted sites on the cellulose)

      Sodium bicarbonate solution, mix for 20 min, repeat twice more.

      Acetate buffer, mix for 1 h. The addition of acetate to the cellulose which has previously been in bicarbonate can lead to a build up of gas in the container. Therefore eye protection should be worn when handling and opening the containers. Repeat with further acetate buffer and leave at 4°C overnight. Repeat next morning, mixing for 1 h.

Protocol 16. *Continued*

Phosphate buffer, mix 20 min.

Protein-phosphate buffer, mix 20 min.

Antibiotic phosphate buffer, mix 20 min.

Phosphate buffer, mix 20 min.

8. Resuspend the solid phase sediment by mixing with phosphate buffer to a final volume of about 800 ml.
9. Test the uniformity of the suspension by drawing it into a 1 ml pipette tip immediately after mixing it by swirling the bottle. If the suspension fills the pipette tip when tested several times, then the dispersion is uniform and can be diluted for use. If the solid phase 'clumps' around the tip, then further dispersion is required by ultrasound. Put the bottle into a sonicator bath and sonicate for 1–2 min. Test for uniformity and if still necessary, mix the suspension by swirling and repeat the sonication. Test and repeat again until a uniform dispersion of the solid phase is reached.
10. Transfer the solid phase suspension to a 2 litre measuring cylinder. Add further phosphate buffer to bring the total volume to 1000 ml (suspension of 150 mg solid phase/ml). Transfer to a 1 litre storage bottle. Rinse out remaining solid phase from the measuring cylinder into the bottle with a volume of the suspension.
11. Store at 4°C.
12. For use in assays resuspend the solid phase by gentle rotation. Add 5 ml of the suspension to 20 ml diluent buffer for solid phases and pipette from this solution (stirred) for use in assays. It is recommended that a new solid phase is added in different amounts to determine optimum assay performance (e.g. 0.5, 1.0, 2.0 mg per tube).

Cyanogen bromide activated alcohol groups have been used for the immobilization of immunoreagents. However, it has been found that the isourea and substituted imidocarbonate bonds formed with amine-containing ligands are subject to a slow but steady dissociation (77).

Tosyl chloride activation (tresyl chloride is similar) converts hydroxyl groups on solid phases to activated sulphonates. Following reaction with nucleophiles such as amines, stable bonds are formed (*Figure 7*) (40,72–74). Methods for the activation and coupling of magnetizable particles Dynabeads™ M-450 using the tosyl chloride method are given (*Protocols 17* and *18*). Dynabeads™ M-450 are uniform, magnetizable polystyrene particles (diameter 4.5 μm).

Protocol 17. Tosyl chloride activation of Dynabeads™ M-450

Materials

- Magnetic particles: Dynabeads™ M-450 (30 mg/ml sterile aqueous suspension)
- Dry acetone

Safety

1. Acetone is highly flammable and should be stored, handled and disposed of with no exposure to flame or spark sources. Disposal into drains can lead to explosion.
2. *p*-Toluenesulphonyl chloride is toxic and should only be handled in a fume hood.

Method

Carry out all procedures in a fume hood.

1. Wash sequentially 0.3 g of Dynabeads™ as follows:

   10 ml water:acetone 7:3

   10 ml water:acetone 6:4

   10 ml water:acetone 2:8

   10 ml dry acetone (three times)

   For washing rotate end over end for 1 min. The washes are removed by aspiration using a Dynal magnetic particle concentrator to sediment the particles.
2. Add 0.60 ml pyridine (BDH, AnalaR) and 0.60 g *p*-toluenesulphonyl chloride (Aldrich) in 8.8 ml dry acetone. Incubate for 24 h at room temperature using end over end rotation.
3. Sediment the particles on the concentrator, remove the solution of *p*-toluenesulphonyl chloride by aspiration.
4. Wash the particles using the reverse sequence of that detailed above (1).
5. Suspend the particles in water and then in 10 ml 1 mmol/litre hydrochloric acid.
6. The activated particles may be stored in this solution for 12 months at 4°C.

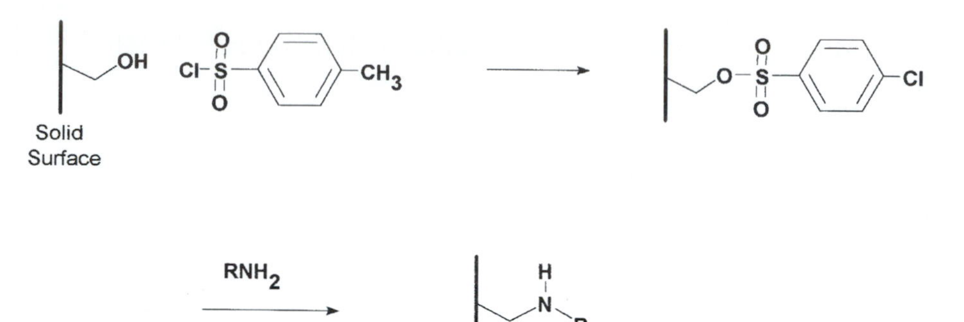

Figure 7. The coupling of protein to hydroxyl groups (e.g. on agarose) activated by *p*-toluenesulphonyl chloride (tosyl chloride). RNH_2 = protein or other amine.

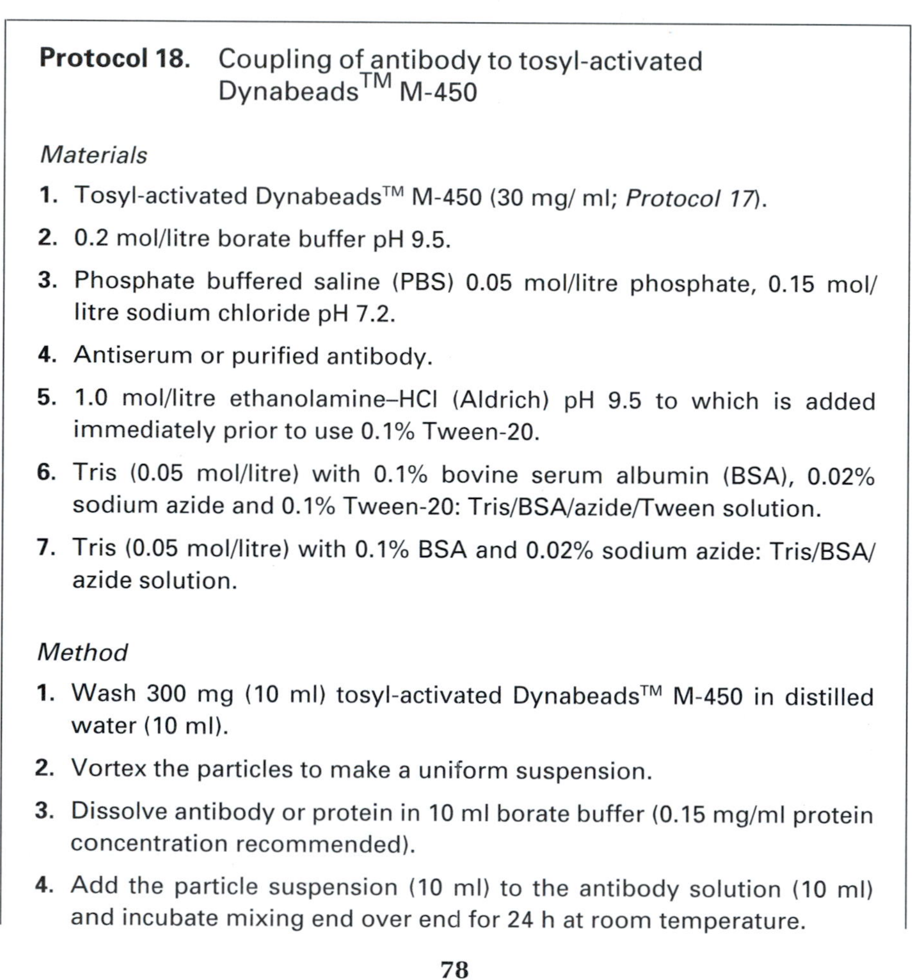

Protocol 18. Coupling of antibody to tosyl-activated Dynabeads™ M-450

Materials

1. Tosyl-activated Dynabeads™ M-450 (30 mg/ ml; *Protocol 17*).
2. 0.2 mol/litre borate buffer pH 9.5.
3. Phosphate buffered saline (PBS) 0.05 mol/litre phosphate, 0.15 mol/ litre sodium chloride pH 7.2.
4. Antiserum or purified antibody.
5. 1.0 mol/litre ethanolamine–HCl (Aldrich) pH 9.5 to which is added immediately prior to use 0.1% Tween-20.
6. Tris (0.05 mol/litre) with 0.1% bovine serum albumin (BSA), 0.02% sodium azide and 0.1% Tween-20: Tris/BSA/azide/Tween solution.
7. Tris (0.05 mol/litre) with 0.1% BSA and 0.02% sodium azide: Tris/BSA/ azide solution.

Method

1. Wash 300 mg (10 ml) tosyl-activated Dynabeads™ M-450 in distilled water (10 ml).
2. Vortex the particles to make a uniform suspension.
3. Dissolve antibody or protein in 10 ml borate buffer (0.15 mg/ml protein concentration recommended).
4. Add the particle suspension (10 ml) to the antibody solution (10 ml) and incubate mixing end over end for 24 h at room temperature.

5. Sediment the particles using Dynal magnetic particle concentrator and aspirate the supernatant.
6. Wash the particles sequentially as follows:

   5 ml PBS for 10 min

   5 ml ethanolamine/Tween solution for 2 h

   5 ml Tris/BSA/azide/Tween solution for 12 h

   5 ml Tris/BSA/azide solution for 2 h

   For washing rotate end over end. The washes are removed by aspiration using a Dynal magnetic particle concentrator to sediment the particles.
7. Sediment the particles and aspirate the supernatant. Resuspend in PBS with 0.1% BSA and 0.02% sodium azide. The beads may be stored for 6 months at 4°C.

Following coupling of the immunoreagent to the solid phase the unreacted sites on the surface need to be blocked. For activated alcohols ethanolamine has often been employed (e.g. *Protocol 16*). Addition of a non-specific protein such as bovine serum albumin (e.g. *Protocol 16*) will couple to unreacted groups as well as block any sites that would passively adsorb protein.

5.2.3 Carboxylic acid functionalized solid surfaces

Methods for the introduction of carboxylic acids onto surfaces have been described earlier (Section 3, *Protocols 3* and *4*; Section 5.2.1, *Protocols 13* and *14*). Solid phase carboxylic acids are commonly activated by reaction with 1-ethyl-3-(3-dimethylaminopropyl)carbodiimide (EDC) (*Figure 8*) or EDC and *N*-hydroxysuccinimide (*Figure 9*) to form active esters (57–60). These groups then react with amines on the molecule to be immobilized forming stable amide bonds. In some cases, it is preferable to carry out the activation of the solid-phase carboxylic acids before the addition of antibody or other protein (*Protocol 22*); this means the protein is not exposed to unreacted EDC which can cause cross-linking and so lead to a loss of immunoreactivity. However, for many antibody preparations it is sufficient to simply add the antibody and EDC to the solid phase in one step (*Protocols 19–21*).

Unreacted active esters may be blocked by the addition of ethanolamine or glycine. Alternatively or additionally a non-specific protein (e.g. bovine serum albumin) may be added to the solid phase. This will block any unreacted groups and also any sites that may bind protein non-covalently.

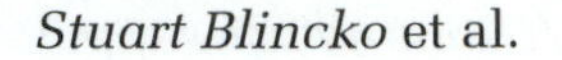
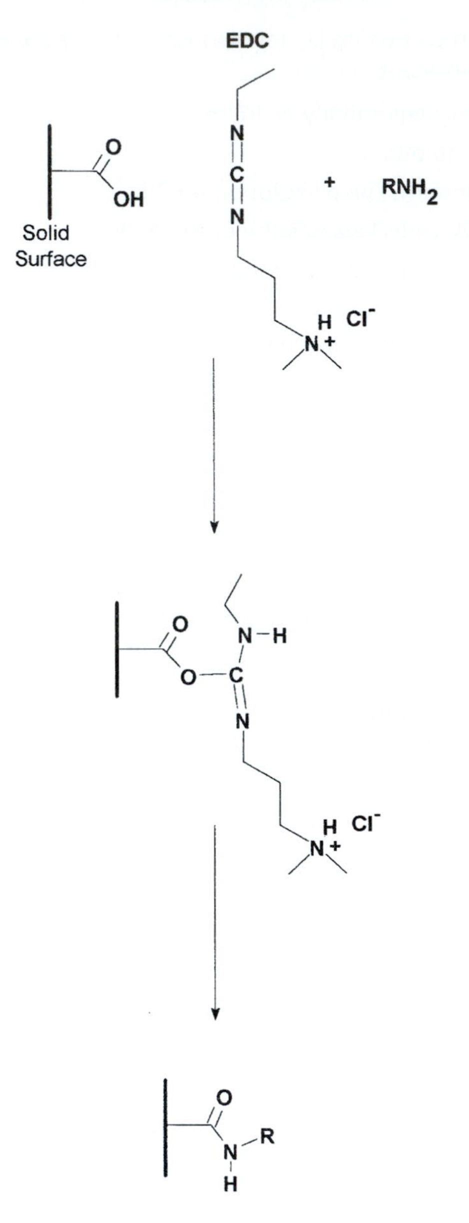

Figure 8. The activation of a carboxylic acid derivatized surface with EDC to form an active ester and its reaction with an amine. The reaction is carried out in one step (cf. *Figure 9*) with all reagents added together. EDC = 1-ethyl-3-(3-dimethylaminopropyl) carbodiimide, RNH_2 = proteins or other amines.

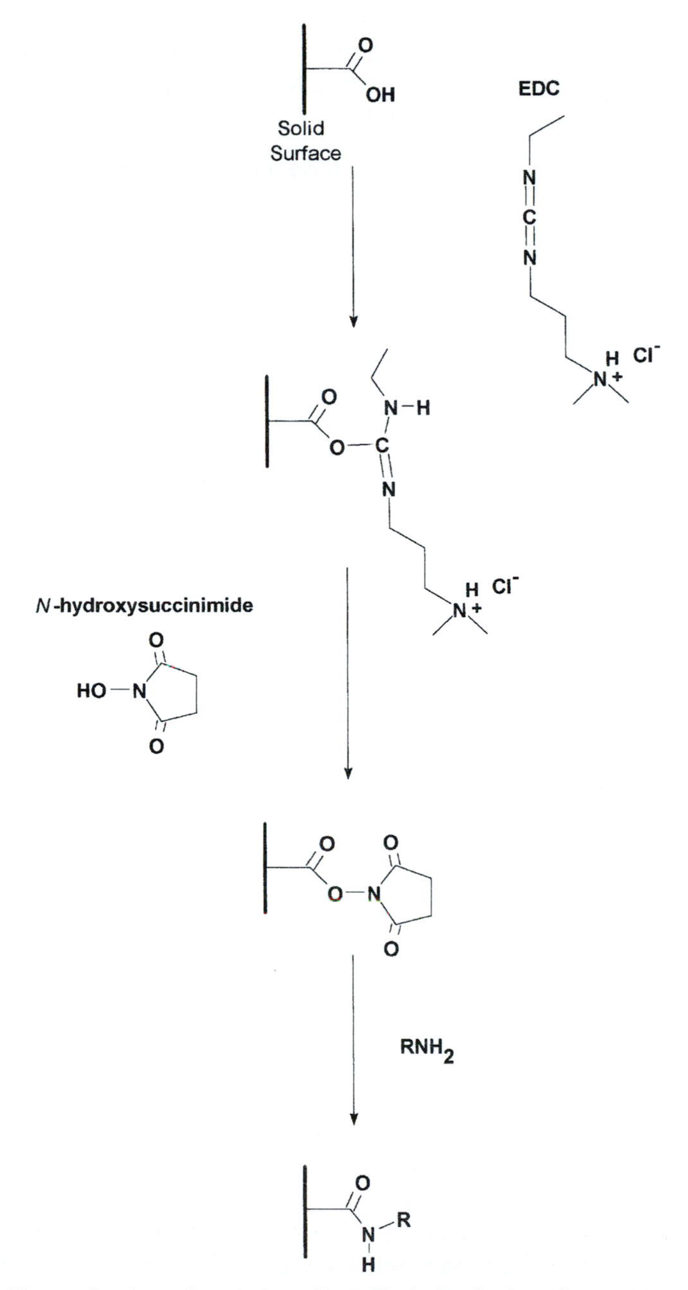

Figure 9. The activation of a carboxylic acid derivatized surface with EDC and *N*-hydroxysuccinimide to form an active ester and its reaction with an amine. The reaction is carried out in two steps, first the activation of the carboxylic acid, then addition of the amine to be coupled. EDC = 1-ethyl-3-(3-dimethylaminopropyl) carbodiimide, RNH_2 = proteins or other amines.

Protocol 19. Covalent coupling of carboxylic acid derivatized magnetizable particles to antibody by the carbodiimide method

1. Take 0.2 g glutaric acid derivatized magnetizable particles (*Protocol 13*) and add 8 ml water.
2. Add 0.2 ml antiserum.
3. Mix for 2 min and add 40 mg EDC. Adjust the pH to 5.6 with 0.1 N NaOH.
4. Rotate the mixture for about 24 h at room temperature.
5. Wash the magnetizable immobilized antibody three times with phosphate buffered saline (PBS, 0.025 mol/litre phosphate, 0.15 mol/litre NaCl, pH 7.4) containing 0.1% bovine serum albumin (BSA) and three times with water, then three times with PBS (30 ml/wash).
6. Resuspend in PBS containing 1% BSA, 0.1% sodium azide and store at 4°C.

Protocol 20. Carbodiimide method for coupling antiserum to polyacrylamide or polystyrene magnetizable particles

The polyacrylamide and polystyrene magnetizable particles had methacrylic acid added to the polymerization mixtures. This gave a carboxylic acid functional group on the surface (*Protocols 3* and *4*).

1. Wash 0.2 g of polyacrylamide or polystyrene magnetizable particles twice with water (30 ml/wash) and resuspend in 0.05 mol/litre 2-(*N*-morpholino)ethane sulphonic acid pH 6.0 (MES) or distilled water.
2. Add 0.2 ml of antiserum. Stir the mixture for 20 min and add 20 mg of EDC.
3. Adjust the pH of the supernatant to 5.6 with 0.1 mol/litre HCl or 0.1 mol/litre NaOH, if necessary.
4. Rotate the contents overnight at room temperature.
5. Wash the magnetizable immobilized antibody obtained twice with phosphate buffered saline (PBS, 0.025 mol/litre phosphate, 0.15 mol/litre NaCl, pH 7.4) containing 0.1% BSA and 0.1% Tween 20, once with water, three times with PBS and once with 0.05 mol/litre Tris buffer pH 7.5 containing 0.004 mol/litre EDTA (30 ml/wash).
6. Resuspend in PBS containing 1% bovine serum albumin (BSA), 0.1% sodium azide and store at 4°C.

Protocol 21. Carbodiimide method for coupling antibody (partially purified) to polyacrylamide or polystyrene magnetizable particles

The polyacrylamide and polystyrene magnetizable particles had methacrylic acid added to the polymerization mixtures. This gave a carboxylic acid functional group on the surface (*Protocols 3* and *4*).

1. Wash 0.2 g of polyacrylamide or polystyrene magnetizable particles twice with water (30 ml/wash) and resuspend in 0.05 mol/litre 2-(*N*-morpholino)ethane sulphonic acid pH 6.0 (MES) or distilled water.
2. Add 2 ml of antibody (caprylic acid precipitation). Stir the mixture for 20 min and add 20 mg of EDC.
3. Adjust the pH of the supernatant to 5.6 with 0.1 mol/litre HCl or 0.1 mol/litre NaOH, if necessary.
4. Complete as for *Protocol 20*.

Protocol 22. Coupling of antibody to polystyrene beads modified with carboxylic acid functional groups on the surface

For the preparation of these beads see *Protocol 14*.

1. Add 400 mg EDC to 10 ml water and add immediately to 30 succinic acid modified beads (*Protocol 14*). Adjust the pH to 9.0 with sodium carbonate. Gently rotate for 2 h at room temperature.

   Optional protocol: add 450 mg *N*-hydroxysulphosuccinimide (Pierce) to the above mixture.
2. Quickly wash the beads in a Buchner funnel with 200 ml cold water (4°C).
3. Immediately add the beads to 50 ml antibody solution. Typically 1–10 mg antibody in 0.1 mol/litre sodium bicarbonate pH 8.0, degassed under vacuum. Gently rotate overnight at 4°C. NB. Avoid preparing the antibody solution with amines or other strong nucleophiles (e.g. Tris or glycine buffers).
4. Wash the beads in a Buchner funnel with 200 ml 0.1 mol/litre sodium bicarbonate and 200 ml water.
5. Add the beads to 50 ml 1% non-specific protein solution (e.g. hydrolysed gelatin or BSA) in 0.1 mol/litre sodium bicarbonate. Gently rotate for three hours at room temperature.

Protocol 22. *Continued*

6. Wash the beads in a Buchner funnel with 200 ml phosphate buffered saline (PBS, 0.025 mol/litre phosphate, 0.15 mol/litre NaCl, pH 7.4) with 0.01% Triton X-100.
7. Add the beads to 10 ml glazing solution (glazing solution = 2% mannitol in distilled water).
8. Aspirate and freeze dry beads.
9. Store the beads with a desiccant in air-tight bags or containers.

5.2.4 Aldehyde-functionalized solid surfaces

Aldehydes couple with amine groups to form Schiff bases. This reaction has been utilized for the coupling of immunoreagents to solid phases using glutaraldehyde. Glutaraldehyde first is bound to amines on the solid surface (Section 5.2.1) and then to amines on the molecules to be immobilized (64,65) (*Figure 10*). To increase the stability of the binding it is usual to reduce the Schiff bases to form substituted amines using sodium cyanoborohydride (*Figure 10*) (*Protocols 23* and *24*). Glutaraldehyde also couples to hydrazide groups on solid surfaces and so enables amine-containing molecules to be covalently bound in an analogous manner. Methods are given for the coupling of antisera to hydrazide-functionalized beads (*Protocol 23*) and amine-functionalized magnetizable particles (*Protocol 24*).

Protocol 23. Coupling of antibody to hydrazide-functionalized polystyrene beads (glutaraldehyde method)

1. Dilute 5 ml of 25% glutaraldehyde to 10 ml in 0.1 mol/litre sodium phosphate pH 7.0.
2. Add 10 ml glutaraldehyde solution to 50 'hydrazide beads' (Pierce). Gently rotate for 2 h at room temperature.
3. Wash the beads in a Buchner funnel (no filter paper) with 200 ml water and 50 ml 0.1 mol/litre sodium phosphate pH 6.0.
4. Dissolve antibody (about 5 mg is most appropriate) in 10 ml 0.1 mol/litre sodium phosphate pH 6.0 and degas under vacuum.
5. Add the glutaraldehyde activated beads to the antibody solution and immediately add 2 mg sodium cyanoborohydride (forms toxic gas—carry out in a fume hood). Gently rotate overnight at room temperature.
6. Wash the beads in a Buchner funnel with 200 ml 0.1 mol/litre sodium phosphate pH 6.0 and 100 ml 0.1 mol/litre sodium bicarbonate.
7. Add the beads to 10 ml of 0.1 mol/litre sodium bicarbonate with 2 mg sodium cyanoborohydride (forms toxic gas—carry out in a fume hood). Gently rotate for 15 min.

8. Wash the beads in a Buchner funnel with 200 ml 0.1 mol/litre sodium bicarbonate and 200 ml water.
9. Add the beads to 10 ml 1% non-specific protein solution (e.g. hydrolysed gelatin or BSA) in 0.1 mol/litre sodium bicarbonate. Gently rotate for 3 h at room temperature.
10. Wash the beads in a Buchner funnel with 200 ml PBS with 0.01% Triton X-100.
11. Add the beads to 10 ml glazing solution (glazing solution = 2% mannitol in distilled water).
12. Aspirate and freeze dry beads.
13. Store the beads with a desiccant in air-tight bags or containers.

Protocol 24. Glutaraldehyde method for coupling antiserum (unpurified) to amino-functionalized magnetic particles

1. Wash 0.2 g of 3-(2-aminoethylamino)propyltrimethoxysilane derivatized magnetic particles (*Protocol 8* or *9*) twice with 0.1 mol/litre phosphate buffer pH 7.4 (PB).
2. Resuspend the particles in 8 ml PB and add 8 ml of 5% glutaraldehyde.
3. Mix for 3 h at room temperature by gently rotating the reaction vessel.
4. Wash five times with PB (30 ml/wash) to remove unreacted glutaraldehyde.
5. Add 0.2 ml unpurified antiserum to the activated particles suspension and rotate for 24 h at room temperature.
6. Wash the antibody coupled to magnetic particles once with PB (30 ml) and resuspend in 15 ml 0.2 M glycine solution.
7. Mix the suspension by shaking for 30 min.

[Following this incubation the Schiff bases may be reduced with sodium cyanoborohydride to give a more stable binding:
   (i) Wash the particles with 0.1 mol/litre sodium bicarbonate.
   (ii) Resuspend the particles in 15 ml 0.1 mol/litre sodium bicarbonate with 2 mg sodium cyanoborohydride. This forms a toxic gas therefore carry out in a fume hood.
   (iii) Rotate for 20 min and then proceed as below.]

8. Wash the magnetizable immobilized antibody with PB, ethanol, twice with phosphate buffered saline (PBS, 0.025 mol/litre phosphate, 0.15 mol/litre NaCl, pH 7.4) containing 0.1% BSA and once with Tris buffer (0.05 mol/litre, pH 7.5) containing 0.004 mol/litre EDTA (30 ml/wash).
9. Resuspend in PBS containing 1% BSA, 0.1% sodium azide and store at 4°C.

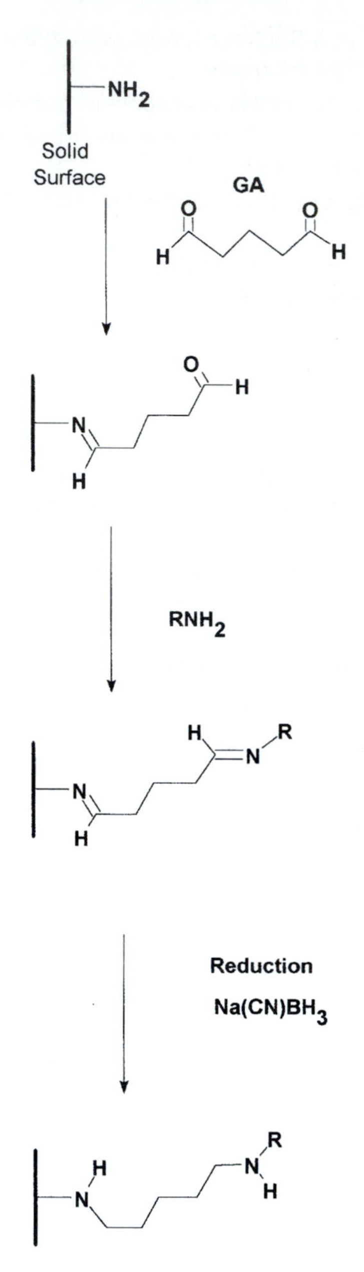

Figure 10. The coupling of amines to an amino-functionalized solid surface using glutaraldehyde (GA). C=N bonds = Schiff bases; sodium cyanoborohydride = $Na(CN)BH_3$; RNH_2 = amine, e.g. protein.

5.2.5 Maleimide-functionalized solid surfaces

Maleimides are added to solid surfaces using heterobifunctional derivatives such as sulphosuccinimidyl-4-(*N*-maleimidomethyl)cyclohexane-1-carboxylate (sulphoSMCC) (67,68). As has been previously described (Section 5.2.1) the *N*-hydroxysuccinimide active ester couples to the amine-derivatized surface to form stable amide bonds. Then the maleimide is reacted with free thiol (sulphydryl) groups on proteins (e.g. antibody fragments such as Fab′) and other molecules to form a thioether bond (*Figure 11*). A method for introducing thiols to proteins is given in Chapter 3, *Protocol 7*. In our experience the maleimide reaction with the thiol should be performed immediately following the coupling of sulphoSMCC to a surface.

Protocol 25. Coupling of antibody fragment (Fab′) to maleimide-functionalized covalink plates

Cold buffers in this protocol (4 °C) minimize the effect of hydrolysis of the active groups.

Materials and buffers

1. Nunc covalink microtitre plates.
2. Sulphosuccinimidyl-4-(*N*-maleimidyomethyl)cyclohexane-1-carboxylate (sulphoSMCC Pierce).
3. Cold (4 °C) phosphate buffered saline (PBS, 0.1 mol/litre phosphate, 0.15 mol/litre NaCl) pH 7.2.
4. Blocking buffer 10 μg/ml cysteine and 0.5% bovine serum albumin in PBS.
5. Washing buffer: PBS + 0.05% Triton X-100.
6. DMSO (BDH, AnalaR).
7. Thiol containing peptide or protein (e.g. Fab′) or a protein derivatized to include a thiol (see *Protocol 7*, Chapter 3).

Method

1. Prepare freshly for use 10 mg/ml sulphoSMCC solution in cold (4 °C) dry DMSO.
2. Dilute the DMSO solution of the sulphoSMCC to a concentration of 250 μg/ml in cold PBS. It is possible to use a lower concentration so a titration is recommended.
3. To as many wells as desired add 100 μl of the sulphoSMCC solution. Cover the wells and incubate at room temperature for 2 h.

Protocol 25. *Continued*

4. Aspirate and quickly wash each well (×2) with cold coupling buffer.
5. Add the thiol-containing peptide or protein (e.g. Fab′) at a concentration of 100 μmol/ ml in cold PBS and incubate at 4°C for 3 h.
6. Wash the wells with washing buffer (x4).
7. Aspirate the wells and add 300 μl blocking buffer. Cover the wells and incubate for 1 h at room temperature.
8. Aspirate and wash each well with washing buffer four times.
9. Store sealed with desiccant at 4 °C.

6. Indirect coupling methods

The methods for covalent binding and passive adsorption may be modified so the primary antibody (the antibody specific for the analyte) is not directly bound to the surface. Instead the surface is coated with molecules that will bind the primary antibody or its derivatives (e.g. streptavidin-coated solid supports will bind biotinylated primary antibodies). This approach enables one solid support preparation to be used for a whole range of primary antibodies.

It has been found that the indirect approach can increase the binding of primary antibody to the surface and improve assay precision.

6.1 Indirect coupling of antibody via a second antibody

Second antibodies are raised against the type and species of the primary antibody (e.g. anti sheep IgG or anti mouse IgG). Obviously the second antibody must be raised in a different species to the primary. Numerous examples include donkey anti sheep IgG and sheep anti mouse IgM. Second antibodies are bound first to a surface (either by passive adsorption or covalent coupling), then the primary antibody is added (78–80) (*Figure 12*). This enables one preparation of solid phase to be used to bind primary antibodies with a variety of specificities (e.g. donkey anti sheep IgG on a solid phase can bind sheep anti thyroxine, sheep anti thyrotropin, sheep anti prolactin etc.). This method can lead to improved assay performance if the results of direct adsorption of the primary antibody are not satisfactory (Section 4).

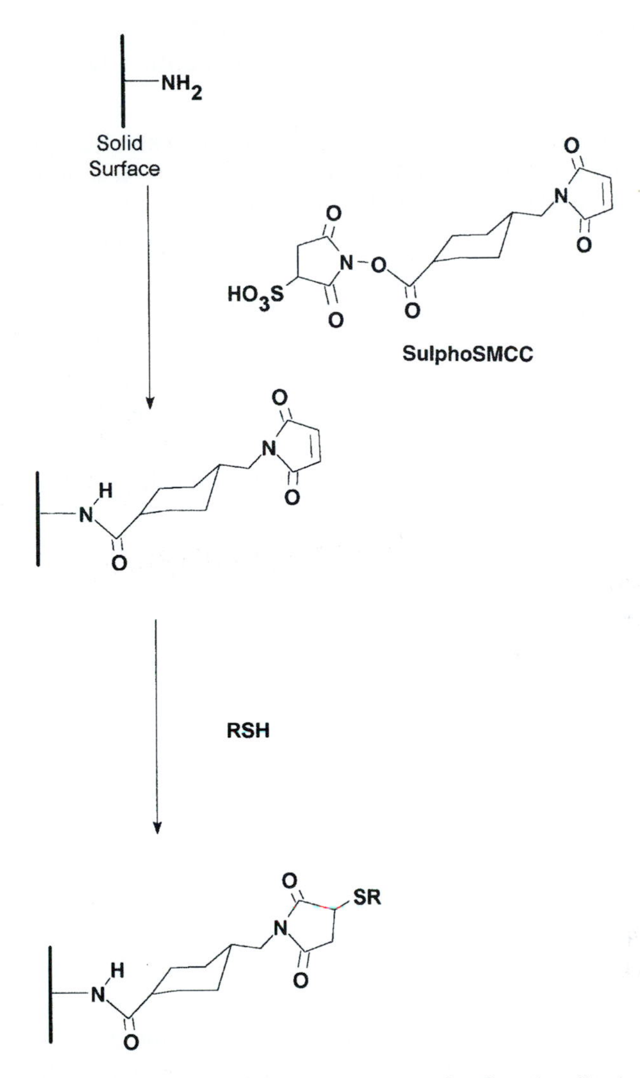

Figure 11. The introduction of maleimide groups to an amine-functionalized solid surface using sulphosuccinimidyl-4-(*N*-maleimidomethyl)cyclohexane-1-carboxylate (sulpho-SMCC) and the coupling of thiol derivatives. RSH = thiol derivative (e.g. Fab′).

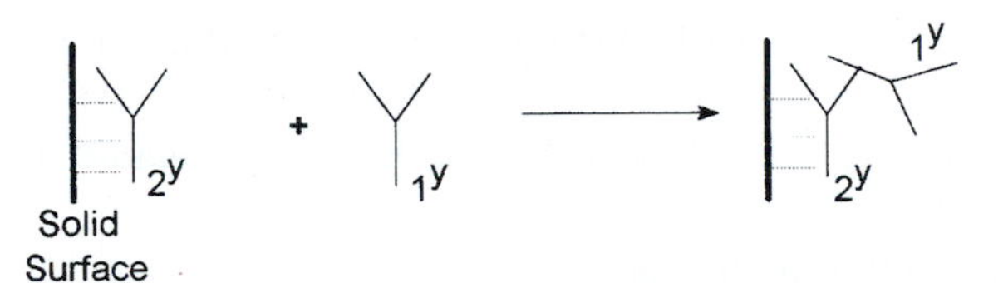

Figure 12. Indirect coupling of antibody via a surface-bound second antibody. Y = antibodies (1^y = primary and 2^y = second antibodies).

Protocol 26. Second antibody binding of antibody to polystyrene microtitre wells

1. Follow the protocol for 'Passive coating of polystyrene microtitre wells with antibody' (*Protocol 5*) using an appropriate second antibody that will specifically bind antibodies from another host animal (e.g. donkey anti sheep serum). A purified form of the second antibody is preferred (e.g. IgG cut).
2. Dilute the primary antibody in phosphate buffered saline (PBS, 0.025 mol/litre phosphate, 0.15 mol/litre NaCl, pH 7.4) with 0.1% BSA and 0.1% sodium azide.
3. Add antibody solution, 200 μl/well, and leave at 25 °C for 24 h in a humid atmosphere (e.g. closed plastic box with wet tissue paper).
4. Aspirate and wash the wells four times with PBS with 0.01% Triton X-100.
5. Add 300 μl/well glazing solution (glazing solution = 2% mannitol in distilled water).
6. Aspirate and vacuum dry the wells.
7. Store the wells with a desiccant in air-tight bags or containers at 4 °C.

Protocol 27. Second antibody binding of antibody to polystyrene beads

1. Follow the protocol for 'Passive coating of polystyrene beads with antibody' (*Protocol 6*) using an appropriate second antibody that will specifically bind antibodies from another host animal (e.g. donkey anti sheep serum). A purified form of the second antibody is preferred (e.g. IgG cut).
2. Dilute the primary antibody in phosphate buffered saline (PBS, 0.025 mol/litre phosphate, 0.15 mol/litre NaCl, pH 7.4) with 0.1% BSA and 0.1% sodium azide.
3. Add antibody solution to cover the beads (e.g. 125 ml to cover 1000 beads in a 250 ml container) and leave closed at 25 °C for 24 h.
4. Aspirate and wash the beads three times with PBS with 0.01% Triton X-100.
5. Add glazing solution (2% mannitol in distilled water) to cover the beads and aspirate.
6. Vacuum dry the wet beads.
7. Store the beads with a desiccant in air-tight bags or containers at 4 °C.

Protocol 28. Second antibody binding of antibody to polystyrene tubes

1. Follow the protocol for 'Passive coating of polystyrene tubes with antibody' (*Protocol 7*) using an appropriate second antibody that will specifically bind antibodies from another host animal (e.g. donkey anti sheep serum). A purified form of the second antibody is preferred (e.g. IgG cut).
2. Dilute the primary antibody in phosphate buffered saline (PBS, 0.025 mol/litre phosphate, 0.15 mol/litre NaCl, pH 7.4) with 0.1% BSA and 0.1% sodium azide.
3. Add the antibody solution to the test tubes (300 μl/tube) and leave at 25°C for 24 h in a humid atmosphere (e.g. a closed plastic box with wet tissue paper).
4. Aspirate and wash the tubes (3 × 2 ml/tube) with PBS with 0.01% Triton X-100.
5. Add glazing solution (2% mannitol in distilled water) 300 μl/tube and aspirate.
6. Vacuum dry the tubes.
7. Store the tubes with a desiccant in air-tight bags or containers at 4°C.

6.2 Indirect coupling of antibody via binding proteins (A and G)

Protein A and Protein G bind specifically to the Fc region of antibodies, especially IgG (81,82). Consequently, surfaces coated with either protein are able to bind antibodies (*Figure 13*). It is thought that this form binding orientates the antibody binding regions away from the solid surface. Protein G will bind all subclasses of mouse, rabbit and goat IgG (as well as most other common species). Protein A binds swine and guinea pig IgG more strongly than protein G, but binds mouse IgG1 and most rat subclasses weakly. For more detailed information on which classes bind contact Pierce for an information sheet and references.

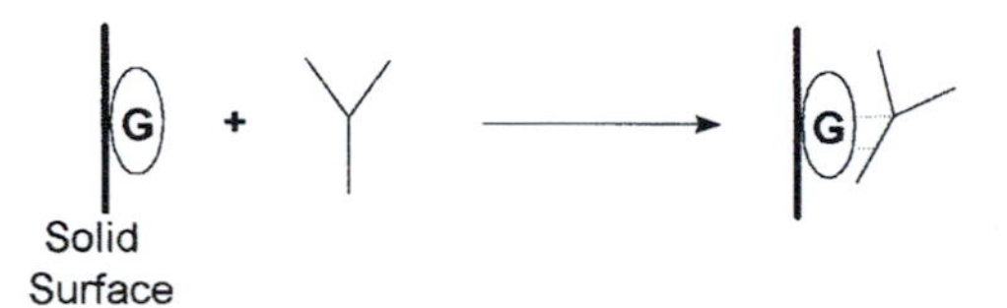

Figure 13. Indirect coupling of antibody to a surface via Protein G. Y = antibody, G = Protein G.

Protocol 29. Coupling of antibody to Protein G coated wells

Materials

1. Reacti-Bind™ Protein G coated wells (Pierce).
2. Antibody able to bind to Protein G (see Section 6.2, above).
3. Coupling buffer: PBS (0.05 mol/litre phosphate, 0.15 mol/litre sodium chloride, pH 7.2) with 0.1% BSA and 0.05% Tween 20.
4. Wash buffer: PBS with 0.05% Tween 20.

Method

1. Dilute the antibody in the coupling buffer (1 mg/litre recommended to begin with). Add 100 μl to each of the Protein G coated wells.
2. Incubate for 1 h at room temperature with mixing.
3. Wash each well four times with 0.3 ml wash buffer.
4. Add 300 μl/well glazing solution (glazing solution = 2% mannitol in distilled water).
5. Aspirate and vacuum dry the wells.
6. Store the wells with a desiccant in air-tight bags or containers at 4 °C.

6.3 Indirect binding of antibody via the biotin–avidin interaction

Avidin or streptavidin bind biotin with a very high affinity (> 10^{15} litre/mol). This has led to a wide range of applications in immunoassay development (83,84), including binding reactions with solid supports. Avidin or streptavidin may be bound to a surface and then a biotinylated primary antibody added. Alternatively, a biotin derivative may be covalently bound to the surface and antibody coupled to avidin or streptavidin added (*Figure 14*). For some applications high NSB has been observed when using avidin and streptavidin possibly due to their basic nature. To help overcome this, more neutral forms of avidin have been made (e.g. NeutrAvidin™ Pierce).

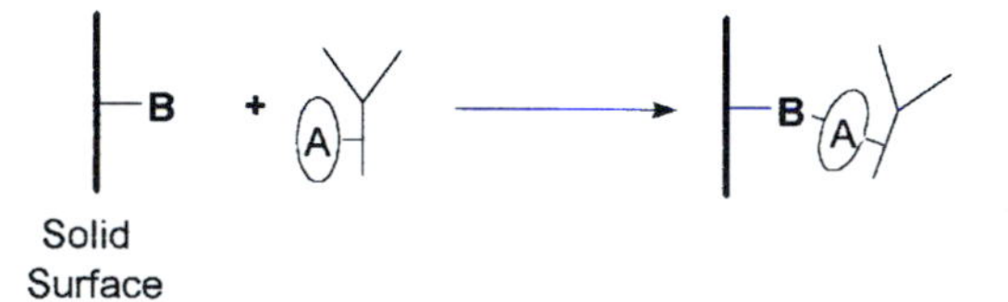

Figure 14. The addition of an antibody–avidin conjugate to a biotin-derivatized solid surface. Y–A = antibody–avidin conjugate, B = biotin bound to the solid support.

Protocol 30. Coupling of antibody–avidin conjugate to biotin-functionalized wells

Materials

1. Biotin-functionalized microtitre wells from Pierce (Reacti-Bind™ Biotin coated wells) or plates prepared by protocol 12.
2. Tris buffered saline (TBS), 25 mmol/litre Tris, 0.15 mol/litre NaCl, pH 7.6.
3. Antibody–avidin conjugate dissolved in TBS with 0.1% BSA and 0.05% Tween-20.
4. Wash buffer: PBS with 0.01% Triton X-100.

Method

1. Add 100 μl antibody–avidin solution to each well. Incubate for 2 h at room temperature.
2. Wash the wells four times with 0.3 ml wash buffer.
3. Add 300 μl/well glazing solution (glazing solution = 2% mannitol in distilled water).
4. Aspirate and vacuum dry the wells.
5. Store the wells with a desiccant in air-tight bags or containers at 4°C.

Protocol 31. Coupling of biotinylated antibody to avidin coated wells

Materials

1. Reacti-Bind™ NeutrAvidin™ coated polystyrene wells (Pierce).
2. Tris buffered saline (TBS), 25 mmol/litre Tris, 0.15 mol/litre NaCl, pH 7.6.
3. Biotinylated antibody dissolved in TBS with 0.1% BSA and 0.05% Tween-20.

Method

1. Add 100 μl biotinylated antibody solution to each well. Incubate for 2 h at room temperature.
2. Wash the wells four times with 0.3 ml wash buffer.
3. Add 300 μl/well glazing solution (glazing solution = 2% mannitol in distilled water).
4. Aspirate and vacuum dry the wells.
5. Store the wells with a desiccant in air-tight bags or containers at 4°C.

6.4 Indirect binding of antibodies labelled with a hapten (fluorescein)

Antibodies raised against a hapten (fluorescein) may be bound first to a surface (e.g. magnetizable particles) and then hapten-labelled primary antibody added (85) (e.g. Serono MAIA clone™; *Figure 15*).

Protocol 32. Indirect binding of antibody labelled with fluorescein to antifluorescein antibodies bound to magnetizable particles

Materials

1. Magnetizable particles coated with anti-fluorescein antibody prepared by one of the protocols above.
2. Fluorescein isothiocyanate labelled primary antibody.

Method

1. Wash 0.2 g antibody coated magnetizable particles in phosphate buffer (0.05 mol/litre).
2. Dissolve the fluorescein-labelled primary antibody in 15 ml phosphate buffer (PB, 0.05 mol/litre phosphate, pH 7.4).
3. Add the primary antibody solution to the magnetizable particles and gently rotate for 24 h.
4. Wash the particles with PB with 0.01% Triton X-100.
5. Resuspend in PB containing 1% BSA and store at 4°C.

7. Blocking

The adsorption or binding of sample components and immunoreagents to the solid surface, apart from the specific antibody antigen interaction, are termed non-specific. Methods to prevent these interactions are essential to achieve the lowest NSB and so to optimize assay performance.

For supports with surfaces capable of passive binding labelled analyte or

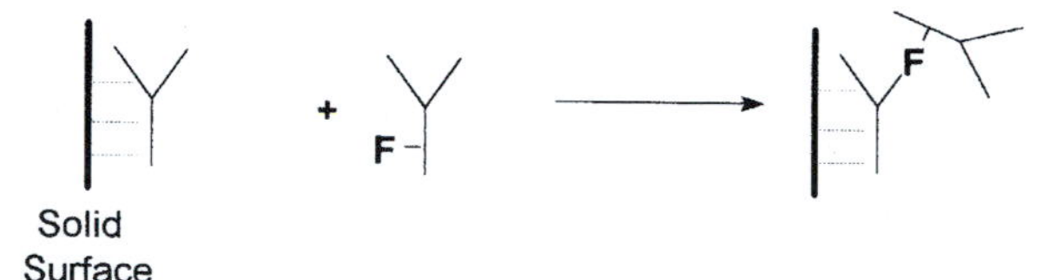

Figure 15. Indirect binding of antibody labelled with fluorescein to antifluorescein antibodies bound to a surface. Y = antibody, Y–F = fluorescein-labelled primary antibody.

other interfering substances it is usual to add a protein or a mixture of proteins to the surface (86). Common blocking agents include dried milk powder, bovine serum albumin, casein, gelatin, hydrolysed gelatin, horse serum and the detergent Tween-20. Preprepared blocking solutions are available from some suppliers (e.g. Pierce, SuperBlock™). Typical protocols add a solution of the above proteins (about 1% w/v) in the same buffer as is used for antibody coating for 1 h at 37°C or overnight at 4°C. Many of the protocols detailed contain instructions for the blocking of the surface to lower NSB.

For supports with activated groups it is necessary to block any unreacted groups after the coupling of antibody or antigen. This is often done by the addition of a small molecule that will react with the unreacted groups but will not lead to an increase in NSB. Such molecules should not be strongly hydrophobic or charged in the assay condition. A good example is ethanolamine which will react with activated groups that couple to amines (e.g. on CDI-activated cellulose). It is not strongly hydrophobic nor is it charged in phosphate buffer pH 7.4. Another approach that may be used as well as or instead of small molecules is the addition of protein or mixtures of protein (as described in the previous paragraph). Proteins will react with unreacted groups on the surface but they will also block any sites that bind proteins non-specifically.

8. Storage

In many situations it has been found that immunoreagents immobilized on solid phase supports are more stable than when in solution (49,50). To gain maximum stability for storage of immunoreagents on dry surfaces (e.g. microtitre wells, beads and tubes) it is recommended that they are stored in air-tight containers (e.g. foil bags) with desiccant at 4°C. Such surfaces may be coated with a sugar solution and vacuum dried for additional robustness (see protocols).

By contrast polysaccharides (e.g. cellulose) and magnetizable particles may be stored in buffered preservative solutions (see protocols).

9. Assay wash solutions

One of the key advantages of solid phase supports is the ease with which they may be removed from the reaction medium and washed. A variety of simple buffered and unbuffered washing solutions have proved useful in immunoassays. Many of the protocols in this chapter employ wash solutions for the preparation of immobilized immunoreagents. Whilst many could be used in immunoassays it is recommended that the following guidelines be taken into consideration.

Table 3. Wash solution constituents

Constituent	**Recommended concentration**
Phosphate (PB)	25 mmol/litre
Phosphate buffered saline (PBS)	25 mmol/litre phosphate, 0.15 mol/litre saline
Tris	50 mmol/litre
Borate	50 mmol/litre
Water only	
Triton X-100	0.01–0.1%
Tween-20	0.05–0.5%
Sodium azide	0.02–0.1% (w/v)

The only way of deciding on which wash solution to adopt is empirically. We have found a buffered wash solution (e.g. PBS with 0.01% Triton X-100) is a good starting point. From there the detergent concentration may be adjusted as well as compared with other detergents (e.g. Tween-20 or 3-[(3-chloramidopropyl)-dimethylammonio]-1-propanesulphonate, CHAPS). High concentrations of detergents (e.g. 0.1% Triton X-100) are more effective at lowering NSB but form a greater amount of foam. Foam is not a problem to assays which are centrifuged but is not acceptable for others, especially where optical measurements are performed. Having determined the detergent type and concentration then a non-buffered solution can be tested. A table of wash solution constituents is given (*Table 3*); it is intended as a guide and not as an exhaustive review.

Wash solutions are generally prepared shortly before use. Where they are to be stored then addition of preservative (e.g. 0.1% sodium azide) is recommended. For assays which are sensitive to the presence of azide (e.g. horseradish peroxidase labelled immunoreagents) then an alternative preservative should be employed.

Appendix A: Suppliers of solid supports

- Microtitre wells for passive coating—Nunc, Costar, Bibby Sterilin, Greiner, Dynatech
- Microtitre wells for covalent coupling—Nunc, Costar, Pierce
- Microtitre wells for indirect coupling (second antibody, avidin, streptavidin, Protein A, Protein G, biotin)—Pierce
- Cellulose, CDI-activated cellulose—Scipac
- Cellulose, agarose, sephacryl and sephadex—Sigma, Pharmacia
- Magnetizable particles—Scipac, Dynal, Advanced magnetics Inc., Reactifs IBF, Bangs

- Test tubes—Greiner, Sarstedt, ICN, Bibby Sterilin
- Beads for passive and covalent coupling—NBL Gene Science, Pierce, Bangs
- Immunofiltration membranes—Pierce
- Organic polymer particles—Bangs

See Appendix B for addresses.

Appendix B: List of suppliers mentioned in this chapter

Aldrich, Gillingham, Dorset, UK

Advanced Magnetics Inc., Cambridge, Massachusetts, 02138, USA

Bangs Laboratories Inc., Carmel, IN 46032–2823, USA

Bibby Sterilin, Stone, Staffordshire, UK

BDH, Poole, Dorset, UK

Costar, Corning Costar Corp., Cambridge, MA02140, USA

Dynal, New Ferry, Wirral, Merseyside, UK

Dynatech, Billingshurst, W. Sussex, UK

Fluka, Gillingham, Dorset, UK

Greiner, GreinerLabortechnik Ltd, Stonehouse, Gloucestershire, UK

ICN, ICN Biomedicals Inc., Costa Mesa, CA92626, USA

NBL Gene Science, Cramlington, Northumberland, UK

Nunc, Kamstrup, DK 4000, Roskilde, Denmark

Pharmacia, St Albans, Hertfordshire, UK

Pierce, Pierce and Warriner, Chester, UK

Reactifs IBF–Pharmindustrie, 92115 Clichy, France

Sarstedt, Leicester, Leicestershire, UK

Scipac, Sandwich, Kent, UK

Sigma, Poole, Dorset, UK

Whatman, Maidstone, Kent, UK

References

1. Ekins, R. P. (1997) In *Principles and practice of immunoassay*, 2nd edn) (ed. C. P. Price and D. J. Newman), pp. 173–207. Macmillan, London.
2. Ekins, R. P. (1983) In *Immunoassays for clinical chemistry* (ed. W. M. Hunter and J. E. T. Corrie), pp. 76–105. Churchill Livingstone, Edinburgh.

3. Wild, D. and Davies, C. (1994) In *The immunoassay handbook* (ed. D. Wild), pp. 57–63. Macmillan, London.
4. Newman, D. J. and Price, C. P. (1997) In *Principles and practice of immunoassay*, 2nd edn (ed. C. P. Price and D. J. Newman), pp. 153–72. Macmillan, London.
5. Self, C. H., Dessi, J. L. and Winger, L. A. (1996) *Clin. Chem.*, **42**, 1527–31.
6. Catt, K. J. and Tregear, G. W. (1967) *Science*, **158**, 1570.
7. Hermanson, G. T., Mallia, A. K. and Smith, P. K. (1992) *Immobilised affinity ligand techniques*. Academic Press, London.
8. Hermanson, G. T. (1996) *Bioconjugate techniques*. Academic Press, London.
9. Butler, J. E. (1991) *Immunochemistry of solid phase immunoassays*. CRC Press, Boca Raton, Florida.
10. Hoyer, L. W. and Trabold, N. C. (1982) *Methods Enzymol.*, **84**, 51–60.
11. Belanger, L. and Masseyeff (1982) *Methods Enzymol.*, **84**, 19–31.
12. Jacobs, D. M. and Gutowski, J. A. (1982) *Methods Enzymol.*, **84**, 264–72.
13. Leigh, B. and Bangs, J. (1990) *J. Clin. Immunoassay*, **13**, 127–31.
14. Hermanson, G. T., Mallia, A. K. and Smith, P. K. (1992) In *Immobilised affinity ligand techniques*, pp. 42–5. Academic Press, London.
15. Douglas, A. S. and Monteith, C. A. (1994) *Clin. Chem.*, **40**, 1833–7.
16. Bangs, L. B. (1984) *Uniform latex particles*. Seradyn Inc., PO Box 1210, IN, USA.
17. Leigh, B. and Bangs, J. (1990) *J. Intl. Fed. Clin. Chem.*, **2**, 188–93.
18. Edwards, R. (1998) Private communication.
19. Dickie, N. and Akhtar, M. (1982) *Methods Enzymol.*, **84**, 254–64.
20. Dakabu, S., Ahene, I. S. and Foli, A. K. (1978) In *Radioimmunoassay and related procedures in medicine, 1977,* Vo1. 1, pp. 155–60. IAEA, Vienna.
21. Crosignani, I. G., Nakamura, R. M., Boyland, D. M. and Mishell, D. R. (1970) *J. Clin. Endocrinol. Metab.*, **30**, 153.
22. Kondorosi, E., Nagy, J. and Denes, G.. (1977) *J. Immunol. Methods*, **16**, 1.
23. Paffard, S. M., Miles, R. J., Clark, C. R. and Price, R. G. (1996) *J. Immunol. Methods*, **192**, 133–6.
24. Hendry, R. M. and Herriman, J. E. (1980) *J. Immunol. Methods*. **35**, 258.
25. Bolton, A. E. and Hunter, W. M. (1973) *Biochim. Biophys. Acta*, **329**, 318–30.
26. Chapman, R. S., Sutherland, R. M. and Ratcliffe, J. G. (1983) In *Immunoassays for clinical chemistry* (ed. W. M. Hunter and J. E. T. Corrie), pp. 178–90. Churchill Livingstone, Edinburgh.
27. Edwards, R.. (1990) *Radioimmunoassay*, pp. 71. IRL Press, Oxford.
28. Ceska, M. (1982) *Methods Enzymol.*, **84**, 238–53.
29. Cowan, S. I., Stagg, B. H. and Niemann, E. (1978) In *Radioimmunoassay and related procedures in medicine, 1977*, Vol. 1, pp. 347–59. IAEA, Vienna.
30. Hermanson, G. T., Mallia, A. K. and Smith, P. K. (1992) In *Immobilised affinity ligand techniques*, pp. 6–9. Academic Press, London.
31. Wide, L. and Porath, J. (1966) *J. Biochim. Biophys. Acta*, **130**, 257–60.
32. Wide, L. (1978) In *Radioimmunoassay and related procedures in medicine, 1977*, Vol. 1, pp. 143–54. IAEA, Vienna.
33. Forrest, G. C. and Rattle, S. J. (1983) In *Immunoassays for clinical chemistry* (ed. W. M. Hunter and J. E. T. Corrie), pp. 147–69. Churchill Livingstone, Edinburgh.
34. Rongsen, S., Ruiyun, X., Fengqi, Z., Xiuzhen, L. and Dingquan, W. (1997) *J. Radioanal. and Nucl. Chem.*, **218**, 131–3.

35. Rongsen, S., Renzhi, W., Ruiyun, X., Yingqi, L., Fengqi, Z., Shaohua, J., Zhihao, L. and Banglei, X. (1996) *J. Radioanal. and Nucl. Chem.*, **206**, 205.
36. Forrest, G. C. (1977) *Ann. Clin. Biochem.*, **14**, 1–11.
37. Nye, L., Forrest, C. G., Greenwood, H., Gardner, J. S., Jay, R., Roberts, J. R. and Landon, J. (1976) *J. Clin. Chim. Acta*, **69**, 387.
38. Guesdon, J. L. and Avrameas, S. (1977) *Immunochemistry*, **14**, 443.
39. Guesdon, J. L., David, B. and Lapeyre, J. (1977) *Ann. Immunol.*, **128C**, 799.
40. McConway, M. G. and Chapman, R. S. (1986) *J. Immunol. Methods*, **95**, 259–66.
41. Johansen, L., Nustad, K., Berg Oerstadvik, T., Ugelstad, J., Berge, A. and Ellingsen, T. (1983) *J. Immunol. Methods*, **59**, 255–64.
42. Neustad, K., Ugelstad, J. and Berge, A. (1982) In *Radioimmunoassay and related procedures in medicine*, pp. 44–55. IAEA, Vienna.
43. Birkmeyer, R. C., Diaco, R., Hutson, D. K., Lau, H. P., Miller, W. K., Neelkantan, N. V. (1987) *Clin. Chem.*, **33**, 1543–7.
44. Ithakissios, D. S. and Kubiatowicz, D. O. (1977) *Clin. Chem.*, **23**, 2072–9.
45. Weetall, H. H. (1970) *Biochem. J.*, **117**, 257.
46. Lim, F. and Buehler, R. J. (1981) *Methods Enzymol.*, **73**, 254–61.
47. Goodfriend, T. L., Ball, D. L. and Updike, S. (1969) *Immunochemistry*, **6**, 481.
48. Donini, S. and Donini, P. (1969) *Acta Endocrinol.*, **63**, Suppl. 142, 257–78.
49. Thakkar, H., Davey, C. L., Madcalf, E. A., Skingle, L., Craig, A. R., Newman, D. J. and Price C. P. (1991) *Clin. Chem.*, **37**, 1248–51.
50. Voller, A. , Bidwell, D. E. and Bartlett, A. (1979) Dynatech Europe, UK
51. Hermanson, G. T., Mallia, A. K. and Smith, P. K. (1992) In *Immobilised affinity ligand techniques*, pp. 9–11. Academic Press, London.
52. Edwards, R.. (1996) Dry surface immunoassays and immunosensors. In *Immunoassays, essential data* (ed. R. Edwards), pp. 96–100. Wiley, Chichester, UK.
53. Weetall, H. H. (1976) *Methods Enzymol.*, **44**, 134–48.
54. Line, W. F. and Becker, M. J. (1975) In *Immobilised enzymes, antigens, antibodies and peptides* (ed. H. H. Weetall), pp. 258. Marcel Dekker, NY.
55. Cuatrecasas, P. (1970) *J. Biol. Chem.*, **245**, 3059.
56. Steers, E., Cuatrecasas, P. and Pollard, H. (1971) *J. Biol. Chem:*, **246**, 196.
57. Hoare, D. G. and Koshland, D. E. (1966) *J. Am. Chem. Soc.*, **88**, 2057.
58. Hermanson, G. T. (1996) In *Bioconjugate techniques*, pp. 144–5 and 170–6. Academic Press, London.
59. Sheehan, J. C., Cruickshank, P. A. and Boshart, G. L. (1961) *J. Org. Chem.*, **26**, 2525–8.
60. Hermanson, G. T. (1996) In *Bioconjugate techniques*, pp. 137–46. Academic Press, London.
61. Wunsch, E. and Drees, F. (1966) *Chem. Ber.*, **99**, 110–20.
62. Staros, J. V. (1982) *Biochemistry*, **21**, 3950–5.
63. Anjaneyulu, P. S. R. and Staros, J. V. (1987) *Int. J. Pept. Protein Res.*, **30**, 117–24.
64. Weston, P. D. and Avrameas, S. (1971) *Biochem. Biophys. Res. Commun.*, **45**, 1574.
65. Ternyck, T. and Avrameas, S. (1972) *FEBS Lett.*, **23**, 24.
66. Klotz, I. M. (1967) Succinylation. *Methods Enzymol.*, **11**, 576.
67. Hermanson, G. T. (1996) In *Bioconjugate techniques*, pp. 235–7. Academic Press, London.
68. Peeters, J. M., Hazendonk, T. G., Beuvery, E. C. and Teeser, G. I. (1989) *J. Immunol. Methods*, **120**, 133–43.

69. Bartling, G. J., Brown, H. D. and Chattopadhyay, S. K. (1973) *Nature*, **243**, 342.
70. Hearn, M. T. W. (1987) *Methods Enzymol.*, **135**, 102.
71. Axen, R., Porath, J. and Ernback, S. (1967) *Nature (London)*, **214**, 1302.
72. Nilsson, K. and Mosbach, K. (1980) *Eur. J. Biochem.*, **112**, 397.
73. Nilsson, K. and Mosbach, K. (1981) *Biochem. Biophys. Res. Commun.*, **102**, 449.
74. Nilsson, K. and Mosbach, K. (1987) *Methods Enzymol.*, **135**, 65.
75. Porath, J. (1974) *Methods Enzymol.*, **34**, 13.
76. Sanderson, C. J. and W.ilson, D. V. (1971) *Immunology*, **20**, 1061–5.
77. Hermanson, G. T., Mallia, A. K. and Smith, P. K. (1992) In *Immobilised affinity ligand techniques*, p. 55. Academic Press, London.
78. Parsons, G. H. (1981) *Methods Enzymol.*, **73**, 224–39.
79. Litt, G. J. (1976) West German Pat. Appl. 2633877.
80. Niswender, G. D. (1977) US Pat. 4048298
81. Akerstrom, B. and Bjorck, L. (1986) *J. Biol. Chem.*, **261**, 10240–7.
82. Schwab, C. and Bosshard, H. R. (1992) *J. Immunol. Methods*, **147**, 125–34.
83. Green, N. M. (1975) *Adv. Protein Chem.*, **29**, 85–133.
84. Diamandis, E. P. and Christopoulos, T. K. (1991) *Clin. Chem.*, **37**, 625.
85. Harmer, I. J. and Samuel, D. (1989) *J. Immunol. Methods*, **122**, 115–21.
86. Hermanson, G. T., Mallia, A. K. and Smith, P. K. (1992) In *Immobilised affinity ligand techniques*, pp. 306–7. Academic Press, London.

3

Enzyme-labelled tests with colorimetric, fluorimetric and chemiluminescent detection systems

IAIN HOWES, STUART BLINCKO, JOHN LITTLE and RAYMOND EDWARDS

1. Introduction

Enzymes are widely used as labels in immunoassay procedures. They demonstrate great versatility and have been applied to all types of immunoassay. Originally employed in immunostaining and gel-precipitin techniques, enzymes subsequently demonstrated particular utility when used with solid-phase separation systems such as antibody coated plastics (e.g. ELISA) and reagent-linked particles (e.g. magnetic cellulose). Enzymes are also used in non-separation assays in which enzyme activity is altered by antibody binding. See Chapter 1 for details of nomenclature and classification.

Enzyme-labelled reagents are used in a variety of formats and are perhaps the most common type of immunodiagnostic test. They are used as detection systems in an increasing number of analysers. This chapter gives information on enzymes suitable for labelling, simple methods for preparation of reagents, testing and storage of labelled reagents and substrates. Assay optimization and solid-phase preparation are detailed in Chapters 1 and 2, respectively.

2. Selection of enzyme-labels

A number of essential criteria must be met for an enzyme to be an effective label. In general terms they are as follows:

1. The enzyme will have a high molar activity (or turnover number).
2. Enzymatic activity can be easily measured.
3. Labelled reagents are immunologically and enzymatically stable.

4. Simple methods for labelling exist which give reproducible products.
5. The enzyme, of a suitable grade, is either easily prepared or is available from commercial sources.

Further points, relating specifically to the intended assay application, may have to be considered. In particular, conditions which will affect enzyme stability and/or activity have bearing on 'signal' generation, for example:

(1) presence of inhibitors;

(2) presence of activators;

(3) substrate analogues;

(4) endogenous anti-enzyme antibodies;

(5) immunostaining;

  (i) labelled reagent has to penetrate the specimen;

  (ii) the product of the enzyme reaction is localized (insoluble);

(6) in non-separation assays the enzymatic activity of the label must be strongly influenced by complex formation with antibody.

In the scientific literature, reference to the use of over 20 enzymes as labels can be found. Only eight are regularly used, the details of which are summarized in *Table 1,* including specific activities of preparations suitable for labelling. Of these eight, two enzymes are found in the vast majority of applications:

- Alkaline phosphatase (AP)
- Horseradish peroxidase (HRP)

They are readily available commercially, may be conjugated by simple techniques and have a range of substrates. To a much lesser extent beta-galactosidase (BG) is used, others only rarely or in special cases, such as lysozyme in homogeneous enzyme immunoassay (EIA). Commonly used combinations of separation methods, instrumentation and enzymes are shown in *Table 2.* Examples of commercial systems and associated technology are given in Chapter 7.

3. Enzyme preparation and assignment of enzymatic activity

3.1 Preparations

Commercial preparations of enzymes are available for labelling immunoreagents. Often termed 'ELISA grade' or 'ELISA purity', they give satisfactory results in most cases and generally do not require further purification. However, in special circumstances or to achieve the highest performance, it may be necessary to repurify an enzyme.

Table 1. Enzyme labels for immunoassay

Enzyme	Abbreviation	Source	M_1	Applications[a]	Specific activity U/mg[b]	Refs
Alkaline phosphates	AP	Bovine intestinal mucosa	84500	EEIA, IEMA, IH, BWB		(23–42)
Alkaline phosphatase	AP	*E. coli*	80000	EIA, IEMA, IH, WB	>7000[c]	
Horseradish peroxidase	HRP	Horseradish roots *Amoracia rusticana*	44000	EIA, IEMA, IH, WB	>200	(11, 43–51)
β-Galactosidase	BG	*E. coli*	116250	EIA, IEMA	>800	(41, 52–55)
Glucose oxidase	GO	*Aspergillus niger*	76500	EIA, IEMA	360	(56)
Urease	–	Jack bean *Canavalia ensiformis*	480000	EIA, IEMA	1400	(57,58, 59)
Lysosyme	–	Egg white	14600	Homogeneous EIA	1500	(60)
Malate dehydrogenase	MDH	Porcine heart mitochondria	35000	Homogeneous EIA	1500	(61,62)
Glucose-6-phosphate dehydrogenase	GPD	*Leuconostoc mesenteroides*	52000	Homogeneous EIA	500	(63,64)
'Microperoxidase'	8-Mp[d]	Proteolysis of horse heart cytochrome c	1502 (8-MP)	IH	N/A	(65)
	9-Mp[d]		1630 (9-MP)			
	11-Mp[d]		1857 (11-MP)			

[a]Classification of immunoassay: EIA— 'limited antibody immunoassay'; IEMA— 'excess antibody immunometric assay; IH—immunohistochemistry; WB—western blot.
[b]Definitions for units of activity (by recommended assays) are given in Section.
[c]*Protocol 7*, Method A.
[d] Number indicates length of peptide.

3.2 Enzymatic assays

Enzyme activity is defined by rate of conversion of substrate to product. Many spectrophotometric methods are in use and the activity observed is dependent upon the test procedure. Specific activity is the enzyme activity per mass unit of protein. In order to make valid comparisons of enzyme preparations from different commercial sources, details of the assay procedure and the method of protein determination (in the case of specific activities) should also be considered and it may be necessary to compare preparations empirically.

Definitions of activity units for enzyme labels are given in *Table 3*. The

Table 2. Common assay formats employing enzyme labels

Immunoreagents	Enzyme substrate and signal generation	
Magnetic particle solid phase	Alkaline phosphatase + dioxetane substrate	Chemiluminescence
	Alkaline phosphatase + 4-MUP substrate	Fluorescence
	Alkaline phosphatase + PMP substrate	Colour
Coated tubes	Horseradish peroxidase + luminol + enhancer	Chemiluminescence
Coated microwells	Horseradish peroxidase + luminol + enhancer	Chemiluminescence
Coated beads	Horseradish peroxidase + TMB	Colour
Coated cartridges	Alkaline phosphatase + 4-MUP	Fluorescence

Table 3. Units of activity

Enzyme	Unit
Alkaline phosphatase	(i) That amount of enzyme causing the hydrolysis of one micromole of *p*-nitrophenyl phosphate (PNP) per minute at pH 9.6 and 25 °C (ii) That amount of enzyme causing the hydrolysis of one micromole of *p*-nitrophenyl phosphate (PNP) per minute at pH 9.8 and 37 °C in the presence of diethanolamine.
Horseradish peroxidase	Guaiacol assay. That amount of enzyme which causes the conversion of one micromole of hydrogen peroxide per minute at 25 °C. Pyrogallol assay. That amount of enzyme which will produce 1 mg purpogallin from pyrogallol in 20 s at pH 6.0 and 20 °C.
β-Galactosidase	That amount of enzyme which causes the hydrolysis of one mole *o*-nitrophenylgalacto- pyranoside (NPG) per minute at pH 7.8 and 37 °C.

NB. Units obtained with alkaline phosphatase method (ii) will be approximately 3.3 times those obtained with alkaline phosphatase method (i).

protocols in this section are recommended for assay of AP, HRP and BG. They should be used for assessment of different preparations of enzyme and enzyme conjugates (see Section 5.3). Use AnalaR or equivalent purity chemicals throughout.

3.2.1 Alkaline phosphatase

Alkaline phosphatase activity is measured directly by the hydrolysis of *p*-nitrophenyl phosphate (1,2).

1. Reaction:

$$p\text{-nitrophenyl phosphate} + H_2O \rightarrow p\text{-nitrophenol} + H_3PO_4$$

2. Unit:

   That amount of enzyme causing the hydrolysis of one micromole of *p*-nitrophenyl phosphate per minute. Buffer, pH and temperature vary in different procedures.

Protocol 1. Alkaline phosphatase assays

Equipment and reagents

- Spectrophotometer, thermostatted
- Glass cuvette, 3 ml, light path 1 cm
- Double distilled water (or equivalent purity) for buffer preparation
- Glycine buffer: 0.25 mM glycine titrated to pH 9.6 with NaOH
- Supplemented glycine buffer: glycine buffer containing 1 mM $MgCl_2$/0.1 mM $ZnCl_2$/10% glycerol
- Magnesium chloride solution: 50 mM $MgCl_2 \cdot 6H_2O$ (10.2 mg/ml)
- 5.5 mM *p*-nitrophenyl phosphate: 2.0 mg/ml in glycine buffer
- Diethanolamine buffer: 1.0 M, titrated to pH 9.8 with conc. HCl, containing 0.5 mM $MgCl_2 \cdot 6H_2O$ (0.1 mg/ml)
- 0.67 M disodium *p*-nitrophenyl phosphate hexahydrate (Sigma): 250 mg/ml in water

Method A—glycine buffer, pH 9.6 (1)

Lambda = 405 nm, $V = 3.00$ ml, $d = 1$ cm, $T = 25$°C, $E^{\mu M} = 18.7$.

1. Put 2.40 ml disodium *p*-nitrophenyl phosphate in the cuvette.
2. Add 0.50 ml 50 mM $MgCl_2 \cdot 6H_2O$.
3. Add 0.1 ml enzyme in dilution in supplemented glycine buffer.
4. Mix immediately and follow the change in absorbance at 405 nm against a blank reaction mixture containing no enzyme.
5. Using the linear part of the curve:

   Specific activity = $(A_{405}/\text{min})/18.7 \times [\text{AP}]$ RM[a]

Method B—diethanolamine buffer (2)

Lambda = 405 nm, $V = 3.10$ ml, $d = 1$ cm, $T = 37$ °C, $E^{\mu M} = 18.2$.

1. Put 3.00 ml diethanolamine buffer in the cuvette.
2. Add 0.05 ml 0.67 M *p*-nitrophenyl phosphate solution and mix.

Protocol 1. *Continued*

3. Equilibrate at 37 °C and add 0.05 ml of enzyme in dilution (0.05–0.06 U/ml AP).
4. Mix and follow change in absorbance at 405 nm against a blank reaction mixture containing no enzyme.
5. Using the linear part of the curve calculate:

   Specific activity = (A_{405}/min)/18.2 × [AP] RM[a]

[a]Protein concentration of AP in the reaction mixture, as mg/ml, determined by Biuret (3,4,5) or using $A_{280\ 1\%}$ = 10.0 with distilled water as diluent.
[b]Units obtained with *Method B* equal approximately 3.3 times those with *Method A*.

3.2.2 Horseradish peroxidase

HRP activity is measured indirectly by the rate of transformation of a hydrogen donor (6). Three assays based on different chromogenic hydrogen donors are commonly used. The method used by Sigma (pyrogallol assay) is not recommended. The Reinheitszahl (R_z) for HRP is defined as the ration of A_{403} to A_{280}, i.e. a measure of protein haemin content. Pure isoenzyme c has an R_z around 3.5. R_z is not a guarantee of specific activity.

1. Reaction:

$$H_2O_2 + DH_2 \xrightarrow{\text{HRP}} 2H_2O + D$$

2. Unit:

   The amount of enzyme which catalyses the conversion of one micromole of hydrogen peroxide per minute at 25 °C.

Protocol 2. Horseradish peroxidase assays

Equipment and reagents

- Glass cuvette, 3 ml, light path 1 cm
- Spectrophotometer, thermostatted
- Double distilled water (or equivalent purity) for all solutions
- Phosphate buffer, PB, pH 7.0: 0.1 M potassium phosphate
- Guaiacol: 2.45 mg/ml in water (Sigma)
- 8 mM hydrogen peroxide: (0.025%),[a] prepare from 30% solution (AristaR, BDH)
- ABTS[R]: 5 mg/ml in PB pH 5.0
- Enzyme diluent: potassium phosphate, pH 6.8, 0.25% Triton X-100
- 9 mM hydrogen peroxide:[a] in water prepare from 30% solution (AristaR, BDH)

Method A—guaiacol (7)

Lambda = 436 nm, T = 25 °C, d = 1 cm, V = 3.00 ml, $E^{\mu M}$ = 25.5.

1. Put 2.80 ml PB, pH 7.0, in the cuvette.
2. Add 0.05 ml guaiacol solution and 0.10 ml enzyme diluted in PB (approximately 0.02 U) and mix.

3. Initiate the reaction with 0.05 ml 8 mM H_2O_2,[b] mixing immediately.
4. Follow the change in absorbance at 436 nm, reading against a blank reaction mixture containing no enzyme.
5. Calculate the rate of change from the linear part of the curve.

$$\text{Specific activity} = \frac{A_{406}/\text{min} \times 4}{25.5 \times [\text{HRP}]^c}$$

Method B—ABTS[R] (8)

Lambda = 405 nm, $V = 3.05$, $d = 1$ cm, $T = 25°C$, $E^{\mu M} = 36.8$.

1. Put 2.90 ml of ABTS[R] solution in the 3 ml cuvette.
2. Add 0.05 ml enzyme in diluent, approx. 0.04–0.06 U/ml, mix and equilibrate at 25°C.
3. When temperature stable initiate the reaction with 0.10 ml of 9 mM H_2O_2 mixing immediately.
4. Read the increase in absorbance at 405 nm against a no-enzyme blank reaction mixture.
5. Calculate the rate of change of absorbance using the linear portion of the curve.

$$\text{Specific activity U/mg} = \frac{A_{405}/\text{min}}{36.8 \times [\text{HRP}]^c}$$

[a]The concentration of H_2O_2 should be checked spectrophotometrically: 8 mM (0.025%) solution in distilled water has $A_{240} = 0.35$.
[b]Four molecules of H_2O_2 are reduced for one molecule of tetraguaiacol produced.
[c]Concentration of enzyme in the reaction mixture (mg/ml) from mass of salt-free powder.

3.2.3 β-Galactosidase

β-Galactosidase activity is measured using the chromogenic substrate 2-nitrophenol-β-D-galactopyranoside (NPDG). Alternative assay procedures include 4-NPDG substrate (9) (4-nitrophenol $E^{\mu M} = 18.5$) and a lactose substrate/galactose dehydrogenase coupled assay in which the formation of NADH is followed spectrophotometrically.

1. Reaction:

$$\text{2-NP-}\beta\text{-D-galactopyranoside} + H_2O \xrightarrow{BG} \text{D-galactose} + \text{2-nitrophenol}$$

2. Unit:

   That amount of enzyme which causes the hydrolysis of 1 μmol of 2-NP-galactopyranoside per minute at 37°C, pH 7.8.

Protocol 3. β-Galactosidase assay (10)

Equipment and reagents

- Spectrophotometer, thermostatted
- Distilled water for all solutions
- Phosphate buffer, PB: 50 mM potassium phosphate, pH 7.8
- Magnesium chloride: 10 mM; 20.3 mg $MgCl_2 \cdot 6H_2O$/10 ml water
- 2-Nitrophenol-β-D-galactopyranoside NPDG: 20 mM; 6 mg/ml in PB
- 2-Mercaptoethanol: 10 M; 0.698 ml up to 1.000 ml with distilled water
- A reference β-galactosidase preparation to check the recovery of the assay

Method

Lambda = 405 nm, V = 2.98 ml, d = 1 cm, T = 37 °C, $E^{\mu M}$ = 3.5.

1. Add 2.2 ml PB, 0.3 ml magnesium chloride, 0.40 ml NPDG and 0.03 ml 2-mercaptoethanol.
2. Mix and allow to equilibrate at 37 °C.
3. Start the reaction with 0.05 ml enzyme dilution (approximately 1.2 U/ml) in PB, mixing immediately.
4. Follow the change in absorbance at 405 nm against a blank solution containing no enzyme.
5. Using the linear portion of the curve calculate:

$$\text{Specific activity} = \frac{A_{405}/\text{min}}{3.5 \times [\text{BG}]^{a}}$$

[a] enzyme protein concentration in the reaction mixture (mg/ml).

4. Conjugation methods

A wide variety of chemical methods have been employed for the linking of enzymes to immunoreagents. The methods can be divided into four sections:

(1) activation of small molecules and their covalent coupling to enzymes;
(2) activation of glycoprotein enzymes and their coupling to proteins;
(3) activation of glycoproteins and their covalent coupling to enzymes;
(4) coupling via bifunctional linking molecules.

4.1 Activation of small molecules and their covalent coupling to enzymes

A wide variety of small molecules have been covalently coupled to enzymes. The methods for their activation and coupling depend on the functional groups on the molecule. In most cases, the functional groups are activated so that they will form covalent bonds with amine groups on the enzyme. These activated groups include isothiocyanates, isocyanates, active esters, mixed anhydrides, sulphonyl halides, epoxides, acid halides and aldehydes (11). For β-galactosidase it is also possible to couple to the abundant thiol groups with maleimide, haloacetyl or pyridyl disulphide activated molecules.

The activation reactions usually involve highly reactive reagents in organic solvents. Once made and purified the activated molecule may then be added to the enzyme for conjugate formation. In general, the aim is to form a 1:1 (mols) molecule:enzyme conjugate. To achieve this a titration of the activated molecule is performed with constant amounts of enzyme. A good starting point is to add a 1.5 molar equivalent of an activated molecule to 1 equivalent of the enzyme. It is possible to measure the incorporation by absorption spectroscopy.

If possible the molecule should be added in aqueous solution. However, for those cases where the activated molecule is not water soluble, addition in DMF or DMSO to the enzyme (in aqueous buffer) may also be successfully employed. It is important in most situations not to exceed an organic solvent: aqueous buffer ratio of 1:20, otherwise the enzyme activity may be seriously impaired.

Once the reaction is complete the conjugate may be purified by gel filtration, concanavalin A chromatography or the use of Protein A.

4.2 Activation of glycoprotein enzymes and their coupling to proteins

Polysaccharide residues on glycoproteins may be oxidized to aldehydes using periodate. Addition of amines to the aldehydes results in covalent Schiff base bond formation. The enzymes HRP and glucose oxidase are glycoproteins and so can be oxidized and coupled to amine-containing molecules (typically IgG or protein) by this method. The Schiff base bonds are unstable to hydrolysis and so are reduced to stable substituted amine groups.

The method is performed in three steps. Firstly the glycoprotein enzyme is oxidized by periodate to afford the aldehyde groups. Then the protein to be conjugated is added. Finally the Schiff base bonds are reduced to substituted amines.

Once the reaction is complete the conjugate may be purified by gel filtration (see *Protocol 8*), concanavalin A chromatography (see *Protocol 9*) or the use of Protein A (see *Protocol 10*).

Protocol 4. Periodate method for HRP: antibody[a] conjugation (13)

Equipment and reagents

- A salt-free HRP preparation, $R_z > 3.2$, specific activity > 200 U/ml (Guaiacol assay)
- HPLC grade water (e.g. HiperSolv, BDH)
- PD-10 G-25 disposable column (Pharmacia)
- 0.1 M sodium bicarbonate: 0.84 g/100 ml
- 8 mM sodium periodate: 1.71 mg/ml in 0.1 M sodium bicarbonate

Protocol 4. *Continued*

A. Activation of HRP

1. Dissolve 5 mg HRP in 0.5 ml 0.1 M $NaHCO_3$.
2. Add 0.5 ml 8 mM $NaIO_4$[b] in 0.1 M $NaHCO_3$.
3. Mix and incubate in the dark for 2 h at 21 °C.

B. Conjugation reaction

1. Rapidly desalt activated HRP using a PD-10 column washed with 0.1 M $NaHCO_3$.
2. Add a portion of activated enzyme to purified antibody prepared in 0.1 M $NaHCO_3$. The weight of HRP should be equal to half that of the antibody, giving a molar ration of 1.5:1.
3. Add sodium cyanoborohydride (as a 0.2 M solution in 0.1 M $NaHCO_3$) to give 5 mM final concentration.
4. Incubate for >4 h[c], in the dark, tightly closed at 20 °C.
5. Purify the conjugate (see Section 5).

[a]The method can be adapted for HRP conjugation of small molecules which contain $-NH_2$, e.g. thyroxine, oligopeptides.
[b]Final concentration 4 mM, optimal for pure isoenzyme c. Other glycoproteins may require higher concentrations of oxidant
[c]May be left overnight.

4.3 Activation of glycoproteins and their covalent coupling to enzymes

The procedure described above (Section 4.2) is also applicable to the coupling of glycoproteins (e.g. many antibodies) to enzymes.

4.4 Coupling via bifunctional linking molecules

This approach involves the use of bifunctional linking reagents. These are molecules that have two reactive groups—one which can bind to the enzyme and the other to the immunoreagent.

4.4.1 The use of homobifunctional reagents

Homobifunctional reagents have two reactive groups that are the same. A typical homobifunctional reagent is glutaraldehyde. It has two reactive aldehyde groups and is therefore able to link any two molecules containing primary amines ($-NH_2$). Efficiency of conjugation is inferior to the hetero-bifunctional reagents (see below), and polymerization is a common problem. The use of a two-step conjugation process will minimize the formation of homopolymers (*Protocol 5*). For a description of the use of glutaraldehyde for the linking of amines see Chapter 2, Section 5.2.4. The Schiff base bonds so formed (aldehyde coupling to an amine) are unstable to hydrolysis and so are reduced to stable substituted amine groups. Sodium cyanoborohydride $Na(CN)BH_3$ is commonly used for this purpose.

Once the reaction is complete the conjugate may be purified by gel filtration (see *Protocol 8*), concanavalin A chromatography (see *Protocol 9*) or the use of Protein A (see *Protocol 10*).

Protocol 5. Two-step glutaraldehyde conjugation of enzyme and antibody

Materials and equipment

- Glutaraldehyde monomer, specially purified for use in electron microscopy (Sigma)
- Sephadex G-25 PD-10 disposable column (Pharmacia)
- Purified enzyme
- Phosphate buffered saline, PBS: 10 mM sodium phosphate, pH 7.4, 150 mM NaCl
- HPLC grade water (e.g. HiperSolv, BDH)

A. Activation of enzyme

1. Incubate 5 mg enzyme (at 50 mg/ml in 0.1 M sodium phosphate, pH 6.8) with 10 μl 25% glutaraldhyde for 18 h in the dark at room temperature.
2. Remove excess glutaraldehyde by rapid desalting with a PD 10 column[a] running in PBS. Collect enzyme in 200 μl.

B. Conjugation

1. Add 2.5 mg antibody[b] in 0.1 M sodium bicarbonate (approx. 1 ml) to the enzyme to give a molar ratio of 1:1.5 antibody to enzyme and mix.
2. Add 30 μl 0.2 M sodium cyanoborohydride in 0.1 M $NaHCO_3$ and mix.
3. Incubate for 24 h at room temperature, in the dark.
4. Purify the conjugate (column chromatography) or, for use in EIH, dialyse versus PBS (three changes of 1000 volumes).

[a]Alternatively prepare a 10 ml sephadex G-25 column.
[b]For example, use 10 mg Ab for AP, 2.5 mg for HRP.

4.4.2 The use of heterobifunctional reagents

Heterobifunctional bifunctional reagents contain two different reactive functional groups. Each group will have a specific reaction chemistry (e.g. one group will react with amines but not thiols, the other will react with thiols and not amines). They have the advantage over homobifunctional linkers in that, with selection of appropriate conditions, homopolymerization of the reactant molecules can be totally avoided. Heterobifunctional linkers of particular use in immunoreagent preparation are shown in *Table 4*. For certain applications

Table 4. Heterobifunctional conjugation reagents

Chemical name	Acronym in common use	Reactive towards	Distance between linked molecules(nm)	Refs
Succinimidyl 4-(*N*-maleimidomethyl)cyclohexane-1-carboxylate[a]	SMCC	$-NH_2$, $-SH$	1.16	(66,67)
m-Maleimidobenzoyl-*N*-hydroxysuccinimide ester[a]	MBS	$-NH_2$, $-SH$	0.99	(78,69)
Succinimidyl 4-(*p*-maleimidophenyl)butyrate[a]	SMPB	$-NH_2$, $-SH$	1.45	(70)
N-Succinimidyl 3-(2-pyridyldithio)propionate[ac]	SPDP	$-NH_2$, $-SH$	0.68	(72)
4-Succinimidyl oxycarbonyl-α-(2-pyridyldithio)toluene	SMPT	$-NH_2$, $-SH$	1.12	(73)
N-Succinimidyl (4-diiodoacetyl)aminobenzoate[a]	SIAB	$-NH_2$, $-SH$	1.06	(74,75)

[a]Require dissolution in a minimal volume of organic solvent before addition to reaction mixture, e.g. dimethyl formamide (DMF). These linkers can be synthesized with a sulphonated *N*-hydroxysuccinimide ester function which improves solubility in aqueous media when conjugating enzyme and antibody. The use of sulphonated analogues may improve yields.

[b]The bridge length can be increased to 1.56 nm by insertion of amidohexanoyl chain.

it may be desirable to vary the distance between the conjugated molecules in order, for example, to prevent steric hindrance or interference with antibody binding.

N-Hydroxysuccinimide esters are utilized in the majority of hetero-bifunctional linkers. Under conditions of mild alkalinity (pH 7.5–8.4) they react with primary amino groups forming a stable amide bond and an *N*-hydroxysuccinimide by-product. All proteins, peptides and many other small molecules contain amino groups suitable for linking.

Sulphydryl (thiol) groups react with maleimides, pyridyl disulphides and acetyl halides. Thiols may be produced from protein disulphide bridges by partial reduction (20 mM dithiothreitol with EDTA or, if reducing antibody, 2-mercaptoethylamine.HCl, which does not cause dissociation of heavy and light chains). Thiol groups may be introduced via primary amino groups using *N*-succinimidyl-*S*-acetylthioacetate (SATA) (*Protocol 7*) or 2-iminothiolane hydrochloride (Trauts' reagent) (71,76).

A typical conjugation of a thiol-containing protein (e.g. antibody fragment Fab′) to enzyme using succinimidyl-4-(*N*-maleimidomethyl)cyclohexane-1-carboxylate (SMCC) proceeds in two steps. Firstly the heterobifunctional reagent is added the enzyme. The *N*-hydroxysuccinimide active ester forms a stable amide bond with amine groups on the enzyme. Secondly the Fab′ is added. The thiol groups couple to maleimide forming a stable thioether. This reaction is described in more detail in Chapter 2, Section 5.2.5 (sulphoSMCC is the sulphonic acid derivative of SMCC and reacts in an identical way).

Protocol 6. Antibody:enzyme conjugation with SMCC (14)

Equipment and reagents

- Use double distilled or HPLC grade water and clean glassware throughout
- Disposable sephadex G-25 PD-10 column (Pharmacia)
- Purified HRP (see *Table 1*)
- Dimethylformamide, DMF: anhydrous dimethylformamide stored under nitrogen (Aldrich 22, 705–6)
- Succinimidyl-4-(*N*-maleimidomethyl)cyclohexane-1-carboxylate (SMCC, Pierce)

Method

1. Dissolve 9 mg HRP in cold PBS to give a concentration of 20 mg/ml.
2. Add 3 mg SMCC in a minimum volume of DMF (no more than 0.025 ml) with gentle stirring.
3. Incubate for 1 h at 4°C.
4. Immediately purify the modified enzyme by gel filtration (G25 sephadex) using cold PBS as the elution buffer. This step removes the unreacted SMCC.
5. Pool the enzyme fractions (HRP is easily seen as a brown colour).

Protocol 6. *Continued*

6. Adjust the modified HRP concentration to about 10 mg/ml (about 0.9 ml).
7. Add 8 mg SATA modified IgG (*Protocol 7*) in 1 ml PBS to the modified HRP solution. A 4:1 molar ratio of enzyme:protein will often give a high specific activity conjugate. Incubate for 2 hours at room temperature.
8. Purify the conjugate (see *Protocols 8–10*).

Protocol 7. Introduction of free thiol to IgG[a] by SATA by the method of Duncan *et al.* (16)

Equipment and reagents

- Phosphate buffer, PB: 50 mM sodium phosphate, pH 7.5, containing 1 mM EDTA prepared in distilled water
- SATA: *N*-succinimidyl-*S*-acetylthioacetate (Pierce) dissolved in anhydrous dimethylformamide stored under nitrogen (Aldrich 22, 705–6) at 4 mg/ml
- IgG prepared in PB at a concentration of 10 mg/ml[b]
- Hydroxylamine: 0.5 M hydroxylamine hydrochloride (Sigma H 2391)/25 mM EDTA in distilled water titrated to pH 7.5 with solid disodium hydrogen phosphate
- 0.1 M sodium bicarbonate, prepared with distilled water

A. Thiolation reaction

1. Add 10 μl SATA in DMF per ml IgG solution[c] with simultaneous vortex mixing.
2. React for 30 min at room temperature.
3. Separate IgG from small molecules by sephadex G-25 gel filtration on PD-10 columns, collecting in PB. This procedure will dilute the protein approx. 1.4 times. For long-term storage freeze the product at $<$20 °C.

B. Deprotection of thiol

1. Add 100 μl hydroxylamine per ml of derivatized protein.
2. Incubate the reaction mixture for 1 h at room temperature.
3. Separate thiol-IgG by PD-10 column chromatography running in 0.1 M sodium bicarbonate.[d]

[a] If thiolating enzyme protein, the optimal concentrations of reactants may be different from those shown for IgG. Start with this protocol as a guide. In the case of HRP, use a 4.4 mg/ml solution of enzyme and 16 mg/ml SATA in DMF.
[b] The substitution efficiency is dependent on IgG concentration which should not be less than 10 mg/ml.
[c] Optimal molar ration of 2.5 SATA:1 IgG, producing an average 1.2 sulphydryl/mol IgG. Never add more than 10 μl organic solvent/ml aqueous solution.
[d] Ready for reaction with SMCC (or other linking reagent) treated enzyme.

5. Purification, storage and testing of enzyme conjugates

Immunoassay results are almost entirely dependent on the quality of enzyme-labelled reagent. For example, sensitivity and precision are adversely affected by the presence of polymerized conjugate which produces raised non-specific signal. A stoichiometry of 1:1 enzyme:immunoreagent is ideal as larger complexes often suffer from loss of enzymatic activity and/or reduced binding, the latter possibly manifesting as reduced affinity. Homopolymers of antibody, antigen or enzyme should be removed. The purification process should also remove the conjugate from free antigen or antibody which has not reacted.

After the labelling reaction is complete proceed as follows:

1. Small molecule (hapten) conjugates:
   (i) remove unreacted molecules from conjugate;
   (ii) separate conjugate from enzyme and polymers (this step often not required).
2. Antibody conjugates:
   (i) separate conjugate and free enzyme, simultaneously resolving polymers;
   (ii) remove enzyme from product.

In practice excellent results can be achieved with simple 'mild' chromatographic techniques, in particular gel filtration. Ion exchange, lectin chromatography, isoelectric focusing and preparative electrophoresis are useful. Affinity chromatography may be used for isolation of hapten–enzyme, but the elution conditions (e.g. acid pH 3–4, chaotropes) can damage the product and are not recommended for antibody conjugates. Protein A can be used for antibody conjugates with elution at pH 4 (*Protocol 10*), Protein G has much higher affinity and elution conditions are harsher (pH 2.7).

Removal of unreacted hapten from conjugated product is simply and efficiently achieved in the majority of cases by simple group separation (desalting) using sephadex G-25 or G-50 (Pharmacia). Prepacked columns, such as PD-10 (Pharmacia) are usually sufficient—use according to the manufacturer's instructions. Optimal conjugation procedures result in little unreacted enzyme remaining in the product and methods such as affinity chromatography or hydrophobic interaction chromatography, to resolve conjugate from free enzyme, are rarely required.

5.1 Gel filtration of conjugate

The following simple protocol generally gives satisfactory results. Modifications may be made according to the properties of the products of interest. The procedure is very applicable to antibody–enzyme conjugates and can also be applied to fractionation of small molecule–enzyme conjugates, particularly if enzyme homopolymers need to be resolved.

Protocol 8. Removal of free enzyme from conjugate by gel filtration

Equipment and reagents

- Use HPLC grade (e.g. HiperSolv, BDH) or double distilled water, degassed, for all solutions
- Chromatographic column, 16 mm x 300 mm[a] (e.g. C16 column, Pharmacia)
- Adjustable peristaltic pump capable of 10–100 ml/h (e.g. pump P-1, Pharmacia)
- UV spectrophotometer with 1 ml UV grade quartz cuvette or flow-through cell
- Sephacryl S-200 HR[b] (approx. 60 ml settled gel, Pharmacia)
- Phosphate buffered saline, PBS: 25 mM sodium phosphate, pH 7.4/150 mM sodium chloride or Tris buffered saline, TBS: 10 mM Tris, pH 7.4/150 mM sodium chloride

Method

1. Wash the sephacyl with buffer[c].
2. Pack the column, pumping at 60 ml/h.
3. Load column[d] and run at 30 ml/h, collecting 2 min fractions (approx. 1 ml).
4. Record A_{280} (and A_{403} for HRP) for each fraction or flow through spectrophotometer during chromatography.
5. Determine which fractions contain enzyme, which antibody and which conjugate. The column may require calibration with molecular weight markers (Sigma).
6. Prepare a pool of fractions containing conjugate, avoiding contamination with free antibody and enzyme as far as possible.[e] If the content of free antibody is low, this preparation may be stored and used for assay purposes.

[a]For HRP conjugates. Increase column length to 60 cm for AP and BG.
[b]For conjugates of HRP (MW 44000 D) and AP (MW 80000 D) use sephacryl S-200 HR; of BG use S-300 HR.
[c]PBS; for AP substitute TBS.
[d]Up to 20 mg protein in up to 2.5 ml (up to 5 ml for 60 cm column).
[e]Resolution of BG conjugates may be poor; alternatively use Protein A chromatography (*Protocol 10*) to remove free BG, followed by dialysis or desalting of the product.

5.2 Lectin chromatography of glycoprotein enzyme conjugates

Lectins bind molecules which contain sugar residues. Concanavalin A reacts with α-D-mannose and α-D-glucose and some stereoisomers. Concanavalin A linked to sepharose can be used to separate glycoprotein from non-glycosylated protein and other molecules. *Protocol 9* is useful for separation and concentration of enzyme conjugate from free antibody or antigen.

Protocol 9. Purification of glycoprotein enzyme conjugates by concanavalin A chromatography[a]

Equipment and reagents

- Chromatographic column, approx. 5 ml
- Concanavalin A sepharose (Pharmacia, product 17–0440–03)
- Phosphate buffered saline, PBS: 25 mM sodium phosphate, pH 7.4/150 mM sodium chloride
- 0.4 M α-D-methylmannopyranoside: 77.6 mg/ml in PBS

Method

1. Prepare a column of 2 ml concanavalin A–sepharose slurry.
2. Wash with PBS (20 ml) under gravity.
3. Load conjugate (e.g. a pool from *Protocol 8*; volume is unimportant, add suitable volumes stepwise. Collect flow-through material.[b]
4. Wash column with PBS (20 ml), adding to flow-through.
5. Elute column with four steps of 1 ml 0.4 M α-D-methylmannopyranoside. Leave column to stand 30 min between each step.
6. The pooled elute (4 ml) should be stabilized and stored appropriately (Section 5.4).

[a]For glycoprotein enzymes, e.g. HRP, GO, Urease. For AP use an alternative method.
[b]Most monoclonal antibodies are non-adherent; polyclonals will contain adherent glycosylated antibody which can be removed by passage on a similar concanavalin A column before conjugation. Free antibody may be reclaimed.

5.3 Fractionation of antibody conjugates by Protein A chromatography

Protein A from *Staphylococcus aureus* (MW = 42000) has a strong affinity for the Fc portion of the IgG molecule. One molecule of Protein A can bind two IgG molecules. However, the strength of binding is dependent on the species and subclass, see *Table 5*, and some IgGs do not bind at all (17). IgG conjugates can be separated and concentrated in one step by Protein A chromatography with the advantage that, in the majority of cases, monoclonal antibody conjugates are eluted by mild acid conditions. Protein G has a higher affinity for a wider range of species IgG. Although an advantage for antibody purification the elution conditions are harsher (pH 2.7) and are not desirable for conjugate purification.

Table 5. Relative binding affinities of Protein A for IgG species

Species	**Protein A**
Mouse	+
Rat	–
Pig	++
Sheep	–
Goat	+
Cow	+
Horse	–
Rabbit	++
Human	++[a]
Dog	++
Chicken	–

++ High affinity; + medium affinity; – insignificant affinity.
[a]IgG_3 is not bound.

Protocol 10. Isolation of antibody conjugate by Protein A chromatography

Equipment and reagents

- Chromatographic column, approx. 5 ml
- Protein A Sepharose Fast Flow (Pharmacia, product 17–0974–01)
- Phosphate buffer, PB: 0.1 M sodium phosphate buffer, pH 8.0
- Citrate buffer: 25 mM sodium phosphate, pH 7.4/150 mM sodium chloride
- PD-10 G-25 sephadex disposable columns

Method

1. Pack 2 ml Protein A Sepharose 4 Fast Flow into the column.
2. Wash the column with sodium citrate buffer (pH 3.0, 5 ml) followed by PB (10 ml).
3. Bring the pH of the impure conjugate solution to 8.0 with Tris base (if necessary).
4. Load the sample at about 5 ml/h (volume independent), collecting 1 ml fractions.
5. Wash with phosphate buffer. The effluent contains non-immunoglobulin protein and non-adherent immunoglobulins.
6. Elute the column with 10 ml 0.1 M sodium citrate buffer.[ab] After elution wash the column with 10 ml phosphate buffer.
7. Monitor protein elution by absorbance at 280 nm.
8. Exchange the eluted fraction with PBS by G-25 sephadex gel filtration using PD-10 columns (load up to 2.5 ml).

[a]Muringe IgG_1 is eluted at pH 6.0, IgG_{2a} at 4.5, IgG_{2b} at 3.5. All antibody can be eluted in a single step with pH 3.0. Most conjugates elute with pH 4.5.
[b]For pH 4.5 and below add 50 μl Tris base (2 M) to rapidly neutralize each fraction.

5.4 Stabilization and storage of enzyme conjugates

Following purification of enzyme conjugates it is essential to store them carefully to preserve their activity. If activators for the enzyme exist, such as metal ions, they should be supplied and inhibitors or inactivators, such as anti-bacterial agents, must be avoided—see *Table 6*. It is always desirable to store conjugates in as high a concentration as possible, especially HRP which is very sensitive to oxidation in dilute solution. Add BSA (1% w/v), especially when freezing is anticipated. If possible, keep stocks under sterile conditions. For enzyme-labelled antibodies the recommended procedure is to add an equal volume of glycerol (AnalaR, BDH) to the final product and keep at –20°C, at which it will not freeze. 50% glycerol also inhibits bacterial growth. Small aliquots can be retrieved without the need to thaw the conjugate. Alternatively add ethylene glycol and keep at 4°C.

In some cases, e.g. labile peptides, the HRP conjugate may not be stabilized by the above procedure. In difficult cases, lyophilize the conjugate with mantel (2% w/v) and BSA (1% w/v) in appropriate quantities (e.g. sufficient for daily use) using a cold-shelf freeze-drier set at –20°C (this is essential if the product contains 150 mM NaCl). Store dried conjugate at <–20°C *in vacuo* or under nitrogen if sealable vials are available.

A number of patent enzyme stabilizers are available commercially, e.g. Protexidase for HRP from ICN or custom preparations from Applied Enzyme Technology Ltd. They allow storage of dilute conjugates at 4°C, or even ambient temperature for extended periods (>1 year).

5.5 Testing of conjugates

When testing conjugates three parameters should be addressed:

- structural integrity and purity
- specific enzymatic activity
- immunologic activity, as evidenced by immunoassay performance

The first item may be examined by chromatographic and electrophoretic methods. The reader is referred to Hames and Rickwood (18) for details of SDS PAGE, blotting and other procedures. Once a conjugation procedure has been adopted and optimized in process control, monitoring of chromatographic profiles during conjugate purification is sufficient.

Specific activity of a conjugate is defined as the enzyme activity per mass unit of immunoreagent. It is a direct measure of the number of molecules of enzyme which are coupled to the antibody or antigen. Protein determination is a significant step in calculating the specific activity in this context. When routinely preparing conjugates it is advisable, though not essential, to assign specific activities to individual preparations. Use assay protocols in Section 3, together with a standard protein assay (e.g. Lowry, biuret).

Table 6. Activators and inhibitors of enzyme labels

Enzyme	**Metal, coenzyme or prosthetic group**	**Activators**	**Inhibitors: AVOID**	**pH optimum**	**Refs**
AP	Zn^{2+} (2/subunit)	Mg^{2+} (0.01 M), diethanolamine[a]	Inorganic phosphate >50 μM, arsenate, EDTA/EGTA/cysteine, amino acids	8–10	(49,77,78)
HRP	Protohaematin IX		NaN_3 >10^{-3} M, O_2, Cl_2, Fl_2, hydroxylamine, styrene, CN >10^{-6} M, hydroxymethylhydrogen peroxide	6–7	(79,80)
BG		Mg^{2+} (2 mM), Mn^{2+} (1 mM)	β-Mercaptoethanol alone, mercurials alone	7.2–7.7	(81)
GO	FAD (one/subunit)		*p*-Chloromercuribenzoate, nitrate, 8-hydroxyquinoline, D-arabinose, 2-deoxy-D-glucose	5–6	(82,83)
Urease		Inorganic phosphate	Univalent cations	4.8	(84,85)
GPD	NAD^+ (one/subunit)	Mg^{2+} (<10 mM)	Divalent cations, phosphate	7.8	(65,86)
MDH	NAD^+ (one/subunit)	Phosphate, arsenate, Zn^{2+}, malate	oxaloacetate, 8-hydroxyquinoline, phenols, sulphite	7.4	(64)

[a]See *Protocol 1.*

The 'acid test' for a new conjugate is performance in assay. In liquid phase or solid phase enzyme immunoassay, the reagent should be titred in an established assay format and compared to existing (control) enzyme-labelled material with regard to:

(1) assay sensitivity;
(2) non-specific binding (in absence of antigen or antibody);
(3) response generation in the assay system (a combination of specific enzymatic activity and immunological activity);
(4) assay precision;
(5) dynamic range (the range of analyte concentrations over which precision is acceptable);
(6) accuracy, using established standard calibrators with reference to target quality control materials, parallelism of sample dilution and recovery of added analyte;
(7) stability over time, with reference to (1)–(6).

In any case, the best performance is achieved at the 'optimum' dilution; this may or may not be satisfactory for the intended assay application. Of course analytical parameters (1)–(6) are functions not only of label but also of antibody concentration/specificity, separation methodology, sample processing and signal detection (reading): see Chapter 1 and Section 6 of this chapter. The test procedure should be appropriately controlled and data must be statistically valid.

Testing of reagents for immunostaining or western blotting is in principle the same: new conjugate is compared with old (if available), having regard to sensitivity of analyte detection.

6. Signal generation

At the conclusion of an enzyme-labelled immunoassay the amount of enzyme label remaining after separation has to be quantified. In non-separation assay systems where the bound and free enzyme has different activity the total enzyme activity requires quantitation. In the majority of cases the appearance of a product of an enzyme-catalysed reaction is monitored.

Three detection technologies have been widely employed: (see Table 12 for detection limits).

1. **Spectrophotometric ('colorimetric')** in which the appearance of a chromophore is measured or judged by eye.
2. **Fluorometric** in which the appearance of a fluorescent product is measured.
3. **Enhanced chemiluminescence** in which the appearance of a luminescent product is measured.

For immunostaining and blotting procedures in which spatial integrity is important the enzyme label and the product must remain localized. To this end insoluble chromophores may be used.

6.1 Spectrophotometric systems for solid phase enzyme-labelled assays

A coloured product is formed from a chromogenic substrate by the enzyme-catalysed reaction. The optical density (OD) is measured in a spectrophotometer at a wavelength corresponding to the absorption maxima (or close to it), and is assumed to follow the Beer–Lambert law:

$$\text{Absorption } A = \varepsilon c l$$

where ε = molar extinction coefficient; c = concentration of chromophore; l = pathlength of measuring cell in cm.

The readings are used directly as assay response. The performance of a typical microtitre plate filter spectrophotometer is shown in *Table 7*, and details of substrates are summarized in *Table 8*.

Table 7. Specifications of typical microplate filter spectrophotometer

Feature	Wavelength (nm)	Limits of error
Accuracy	405	$<\pm 1\%$ and ± 0.005 OD at 1 OD
Linearity	340–399	$<\pm 0.7\%$ and ± 0.005 OD from 0.1–2 OD
	400–750	$<\pm 0.5\%$ and ± 0.005 OD from 0.1–3 OD
Repeatability	340–399	$<\pm 0.5\%$ and ± 0.005 OD at 1 OD
	400–750	$<\pm 0.3\%$ and ± 0.005 OD

Table 8. Chromogenic substrates for separation EIA

Enzyme	Substrate	Acronym	Wavelength (nm)	Refs
HRP	3,3′,5,5′-Tetramethylbenzidine	TMB	450[a]	(87)
HRP	*o*-Phenylenediamine[b]	OPD	492[a]	(88)
HRP	*o*-Dianisidine[b]	–	400	(89)
HRP	2,2′-Azino-di-(3-ethylbenzthiazoline-6-sulphonate)	ABTS	415[c]	(90)
HRP	5-Aminosalicylic acid	5AS	492[c]	(91)
AP	Phenolphthalein monophosphate[b]	PMP	550	(92)
AP	*p*-Nitrophenyl phosphate	PNP	405	(93)
AP	1-Napthyl phosphate	1NP	405	(94)
Urease	Bromocresol purple/urea	–	590	(95)
BG	*o*-Nitrophenyl-β-D-galactopyranoside	NPG	405	(96)

[a] Stop reaction with H_2SO_4 (final concentration 0.5 M).
[b] Mutagenic or carcinogenic—use with care.
[c] Reaction can be stopped with cyanide (5 mM final concentration) CARE!

6.1.1 End-point determinations

The protocols given below are suitable for the majority of enzyme-labelled assays and are recommended for spectrophotometric detection. A single reading is taken (with or without an 'off peak' reference reading) and used as the assay response. 'Stopping' the signal generation by killing the enzyme may be done, e.g. with 0.5 M acid. Since most readers are rapid (1–5 s for a 96 well plate) stopping is not essential. Stop reagents may change the wavelength and intensity of maximum absorbance—see TMB *Protocol 11*.

Protocol 11. TMB substrate system—for use in HRP-labelled assays (19)

Reagents

- Use HPLC grade (e.g. HiperSolve, BDH) or double distilled water, degassed, for buffer
- Tetramethylbenzidene, TMB stock solution: 10 mg/ml (Sigma product T.2885) in dimethylsulphoxide (SpectrosoL, BDH)
- 0.1 M sodium acetate buffer: 1.36 g sodium acetate trihydrate (AnalaR, BDH) in 100 ml water
- Titrate to pH 6.0 with citric acid
- H_2O_2: peroxide 30% (w/v) solution (AristaR, BDH)

Method

1. Add H_2O_2 to sodium acetate buffer (15 μl per 100 ml, giving 1.3 mM).
2. Just before use add stock TMB to acetate buffer (1 part per 100, giving a final concentration of 0.1 mg/ml) and mix immediately.
3. Dispense required volume to assay system, e.g. 100 or 200 μl per well of microtitre plate.
4. Incubate for 5–30 min[a] with either intermittent (every 60 s) or continual mixing.
5. Stop the reaction with 0.25 volume (e.g. 25 or 50 μl per microwell) 2.5 M H_2SO_4 (BDH, volumetric grade), mixing immediately.
6. Read absorbance at 450 nm with a reference at 650 nm in a suitable spectrophotometer.[b]

[a]Do not exceed 1.5 AU at 650 nm.
[b]Do not exceed 3.0 AU at 450 nm.

Protocol 12. PMP substrate system—for use in AP-labelled assays (20)

Reagents

- Use HPLC grade (e.g. HiperSolve, BDH) or double distilled water, degassed, to prepare all solutions
- DEA buffer: 10% (w/v) diethanolamine (AnalaR, BDH), titrated to 9.8 with 1 M HCl (AnalaR, BDH).
- Add $MgCl_2{\cdot}6H_2O$ (100 mg/ml) and NaN_3 (200 mg/litre)
- PMP:[a] phenolphthalein monophosphate (P 5758, Sigma) or monopyridine PMP (P 5883, Sigma)

Protocol 12. *Continued*

Method

1. Just before required dissolve PMP[a] in DEA buffer to give 1.5 mg/ml.
2. Dispense the required volume to the assay system, e.g. 100 or 200 μl per well of a microtitre plate
3. Incubate for 5–30 min.[b]
4. Read at 550 nm in a suitable spectrophotometer.[c]

[a]Can substitute *p*-nitrophenyl phosphate (PNP).
[b]Do not exceed 3.0 AU at 550 nm.
[c]Reaction need not be stopped. If stopping is desired add 0.25 times substrate volume (e.g. 25 or 50 ml per microwell) of diethanolamine buffer containing 2 M NaOH (AnalaR) and 0.2 M EDTA (AnalaR).

Protocol 13. NPG substrate system—for use in BG-labelled assays (21)

Reagents

- Use HPLC grade (HiperSolv, BDH) or double distilled water for the buffers
- Phosphate buffer, PB: 0.1 M sodium phosphate buffer pH 7.0, containing magnesium chloride (2 mM), manganese sulphate (1 mM) and β-mercaptoethanol (0.6 mM)
- 2-Nitrophenyl-D-galactopyranoside, NPG (Sigma product N1 127)
- Stop solution: PB containing 2 M NaOH (AnalaR, BDH)

Method

1. Just before use dissolve NPG (Sigma) PB to give 2.7 mM.
2. Add the required volume to the assay system (e.g. 100 or 200 μl per well in the microtitre plate system).
3. Incubate for 5–30 min.[a]
4. Stop the reaction with 0.25 volumes stop solution (e.g. 25 or 50 μl per well).
5. Read the absorbance at 405 nm in a suitable spectrophotometer.

[a]Do not exceed 3.0 AU at 405 nm.

6.1.2 Kinetic reading

Absorbance readings are taken as the enzyme-catalysed reaction proceeds and related to time. The rate of change in absorbance is calculated and used as the response. Readers and software which can calculate the reaction rates are widely available, e.g. V_{max} (Molecular Devices) and Multicalc (Pharmacia Wallac).

6.2 Enzyme amplification

Essentially two enzyme systems are coupled, the product from the first representing the substrate for the second, cycling may occur and these systems offer potentially enhanced sensitivity, although they can be prone to interferences. For extensive details see Chapter 6.

6.3 Immunostaining procedures

Substrates which produce coloured insoluble products suitable for immunostaining and western blotting procedures are given in *Table 9*. See references for details of use.

6.4 Fluorescent systems

Fluorogenic substrates offer potential advantages in terms of detection limit. This may be due to the low background signal from the sample and/or the opportunity for repeated measurements.

Ideally the product is highly fluorescent with optical properties significantly different from those of the substrate. A large Stock's shift (the difference between excitation and emission wavelengths) is desirable. Fluorogenic substrates are available for horseradish peroxidase, alkaline phosphatase and β-galactosidase, they are listed in *Table 10*.

Table 9. Substrates for immunostaining (insoluble products)

Enzyme	Substrate	Acronym	Refs
AP	5-Bromo-4-chloro-3-indolyl phosphate	BCIP	(97)
AP	1-Napthyl phosphate	1NP	(98)
HRP/microperoxidase	3,3′-Diaminobenzidine[a]	DAB	(99)
HRP/microperoxidase	4-Chloro-1-napthol	4C1N	(100)
HRP/microperoxidase	3-Amino-9-ethylcarbazole	AEC	(101)
BG	*o*-Nitrophenyl-β-D-galactopyranoside	NPG	(96)
BG	5-Bromo-4-chloro-3-indolyl-β-D-galactopyranoside	BCIG	(102)
BG	Napthol AS-BI-β-D-galactopyranoside	–	(103)

[a]Mutagenic or carcinogenic—use with care.

Table 10. Flurogenic substrates

Enzyme	Substrate	Reference
Alkaline phosphatase	4-Methylumbelliferone phosphate (4-MUP)	(104)
Alkaline phosphatase	Attophos™	(105)
Horseradish peroxidase	3-*p*-Hydroxyphenylpropionic acid (HPPA)	(106)
Horseradish peroxidase	QuantaBlu™	(107)
β-Galactosidase	Resorufin-β-D-galactopyranoside	(108)

6.5 Enhanced chemiluminescent systems

Chemiluminescent substrates offer the advantage of shorter signal generation time, i.e. the incubation of substrate with enzyme may be reduced. Chemical enhancers may be used to boost the light emission up to 100-fold. There are two basic enhanced chemiluminescent systems potentially available for enzyme immunoassay use. Full details of the substrates and their enhancers are not in the public domain and represent patented, confidential or restricted information. The two systems are exemplified by the Amerlite system (Johnson and Johnson) based around horseradish peroxidase, luminol and enhancers: and the Immulite system (DPC Inc.) based around alkaline phosphatase, adamantyl dioxetane phosphate and enhancers. Commercially available substrates are give in *Table 11*.

Table 11. Luminogenic substrates

Enzyme	Substrate	Reference
Alkaline phosphatase	Elisa Light™	(109–118)
Alkaline phosphatase	Lumi Phos® 530	(119)
Horseradish peroxidase	ECL™	(120–133)
Horseradish peroxidase	SuperSignal® LBA Pierce	(134)

Table 12. Detection limits for various labels

Type of label	Detection limit (zeptomoles, 10^{-21} M)
Alkaline phosphatase	
Tropix dioxetane chemiluminescence	1
Fluorescence	100
Colour	50000
Horseradish peroxidase	
Luminol chemiluminescence	25000
Colour	2000000

Protocol 14. 4-Methylumbelliferylphosphate (4-MUP): fluorogenic substrate for alkaline phosphatase

Reagents

1. Buffer (1.0 M diethanolamine, pH 9.0, containing $MgCl_2$):
   (i) Dissolve 10.5 g diethanolamine (BDH Analar 10393) in 100 ml HiperSolv or equivalent H_2O (BDH Analar 15273) adjust to pH 9.0 with concentrated HCl (BDH Analar 10125) prior to making up to volume.
   (ii) Add 10 mg $MgCl_2 \cdot 6H_2O$ (BDH Analar 101494).

2. To prepare working substrate solution dissolve 4-methylumbelliferyl phosphate (JBL Scientific Cat. 1150) 1.0 mg/ml in the diethanolamine/ $MgCl_2$ buffer.

Method

1. Add 100 μl of working substrate solution per well.
2. Incubate for 5–30 min at room temperature.
3. Stop the reaction by the addition of 100 μl of 1.0 M sodium hydroxide.
4. Read relative fluorescent units (RFUs) in a fluorimeter:

   $\lambda_{excitation} = 360$ nm

   $\lambda_{emission} = 450$ nm

Protocol 15. JBL Scientific AttoPhos™: fluorogenic substrate for alkaline phosphatase

Reagents

1. Buffer (diethanolamine pH 9.2):
   (i) Dissolve 26.25 g diethanolamine (BDH Analar 10393) in 100 ml HiperSolv or equivalent H_2O (BDH Analar 15273) adjust pH to 9.2 using concentrated HCl (BDH Analar10125) prior to making up to volume.
   (ii) Add 4.7 mg $MgCl_2{\cdot}6H_2O$ (BDH Analar 101494).
   (iii) Add 5 mg sodium azide (BDH Analar 103094).
2. To prepare working substrate solution dissolve 0.58 mg/ml AttoPhos™ in the diethanolamine buffer.

Method

1. Following standard ELISA procedures, incubate, block and wash the microtitre plate wells as required.
2. Add 200 μl working solution per well and incubate for 30 min at 37 °C.
3. Read relative fluorescent units (RFUs) in a fluorimeter:

   $\lambda_{excitation} = 430–440$ nm

   $\lambda_{emission} = 550–560$ nm

   The AttoPhos™ reagents are now available in kit form from JBL Scientific Inc.

Protocol 16. 3-(*p*-Hydroxyphenyl)-propionic acid (HPPA): fluorogenic substrate for horseradish peroxidase

Reagents

1. Tris buffer 0.1 M pH 9.0:
   (i) Dissolve 1.21 g Tris (BDH Analar 10315) in 100 ml HiperSolv or equivalent H_2O (BDH Analar 15273) adjust to pH 9.0 with concentrated HCl (BDH Analar 10125) prior to making up to volume.
   (ii) Add 0.75 g HPPA (Fluka 56190): this is solution A.
2. Citrate–acetate buffer
   (i) Dissolve 0.272 g sodium acetate (BDH Analar 10235) in 100 ml HiperSolv or equivalent H_2O (BDH Analar15273). Adjust pH to 6.0 by the addition of solid citric acid (BDH Analar 10081).
   (ii) Add 200 μl 30% peroxide (Sigma H1009): this is solution B.
3. To prepare the working substrate solution mix A(5 volumes) and B(1 volume) prior to use.
4. Stopping solution, 0.75 M glycine pH 10.3:
   (i) Dissolve 5.63 g glycine (BDH Analar 101196) in 100 ml HiperSolv or equivalent H_2O (BDH Analar15273).

Method

1. Following standard ELISA procedures, incubate, block and wash the microtitre plate wells as required.
2. Add 100 μl per well HPPA working solution.
3. Incubate at ambient temperature for 30–60 min.
4. Stop the reaction by addition of 100 μl per well of glycine stopping solution.
5. Read relative fluorescent units (RFUs) in a fluorimeter:

$$\lambda_{excitation} = 320 \text{ nm}$$

$$\lambda_{emission} = 404 \text{ nm}$$

Protocol 17. Pierce QuantaBlu™ fluorogenic peroxidase substrate

Contents

- QuantaBlu™ substrate solution
- QuantaBlu™ stable peroxide solution
- QuantaBlu™ stop solution

Preparation of QuantaBlu^{TM} working solution

Mix 9 parts QuantaBlu^{TM} substrate solution to 1 part of QuantaBlu^{TM} stable peroxide solution prior to use. This working solution is stable for a minimum of 24 hours at room temperature and is not light sensitive.

Method

1. Following standard ELISA procedures, incubate, block and wash the microtitre plate wells as required.
2. Add 100 μl of QuantaBlu^{TM} working solution to each well.
3. Incubate for 5–90 min either at room temperature or at 37 °C. Note for prolonged incubation times or incubation at elevated temperatures cover the wells with sealing tape to prevent evaporation.
4. Stop the peroxidase activity by the addition of 100 μl of QuantaBlu^{TM} stop solution. The enzyme activity is immediately stopped and no further incubation is required.
5. Read relative fluorescent units by fluorimetry:

   $\lambda_{excitation} = 325$ nm

   $\lambda_{emission} = 420$ nm

Protocol 18. Resorufin-β-D-galactopyranoside: fluorogenic substrate for β-galactosidase

Preparation of working substrate solution

Dissolve 75 mg/l resorufin-β-D-galactopyranoside in 10 mM potassium phosphate buffer, pH 7.5, containing 150 mM NaCl and 2.0 mg $MgCl_2$.

Method

1. Following standard ELISA procedures, incubate, block and wash the microtitre plate wells as required.
2. Add 200 μl working substrate solution.
3. Incubate for 30 minutes at 37 °C.
4. Read relative fluorescent units (RFUs) in a fluorimeter:

   $\lambda_{excitation} = 550$–$572$ nm

   $\lambda_{emission} = 583$ nm

Protocol 19. Tropix ELISA LIGHT^{TM}: chemiluminogenic substrate for alkaline phosphatase

Method

1. Following standard ELISA procedures, incubate, block and wash the microtitre plate wells as required.
2. Add 100 μl working solution per well and incubate for 5–10 minutes at 37 °C.
3. Measure chemiluminescence produced in a luminometer. The emission wavelength will depend upon the enhancer employed in the system:

   Sapphire II $\lambda_{emission}$ = 463 nm

   Emerald II $\lambda_{emission}$ = 542 nm

Protocol 20. Lumigen Inc. Lumi-Phos^{R} 530: chemiluminogenic substrate for alkaline phosphatase

Reagents supplied

- Ready to use solution containing Lumigen^{TM} PPD
- 4-Methoxy-4(3-phosphatephenyl)-spiro(1,2-dioxetane-3,2′-adamantane) disodium salt 0.33 M
- Fluorescein surfactant 0.035 M
- 2-Amino-2-methyl-1-propanol buffer, pH 9.6, 0.75 M
- $MgCl_2$ 0.88 M
- Cetyltrimethyl ammonium bromide 1.13 mM

Method

1. Following standard ELISA procedures, incubate, block and wash the microtitre plate wells as required.
2. Add 200 μl working solution per well and incubate for 30 min at 37 °C.
3. Read luminescence generated in a luminometer:

   $\lambda_{emission}$ = 477 nm

Protocol 21. ECL Amersham: chemiluminogenic substrate for Horseradish peroxidase

Contents

- Amerlite signal reagent buffer
- Tablet A
- Tablet B

Method

1. Following standard ELISA procedures, incubate, block and wash the microtitre plate wells as required.
2. Prepare the working substrate solution by the addition of one tablet A and one tablet B to the Amerlite signal reagent buffer prior to use. This working reagent is stable at room temperature for several hours and is not light sensitive.
3. Add 250 μl working substrate solution per well and incubate at 25°C for 2–20 min.
4. Read the luminescence generated in a luminometer at $\lambda_{emission}$ = 420 nm.

Protocol 22. Pierce SuperSignal^{R} LBA: chemiluminogenic substrate for horseradish peroxidase

Contents

- Luminol/enhancer LBA
- Stable peroxide LBA

Preparation of SuperSignalR LBA substrate working solution

Mix equal parts luminol/enhancer LBA and stable peroxide LBA prior to use. This working solution is stable for a minimum of 24 h at room temperature and is not light sensitive.

Method

1. Following standard ELISA procedures, incubate, block and wash the microtitre plate wells as required.
2. Add 100–150 μl SuperSignal^{R} working solution to each well.
3. Mix liquids in wells for 5 min using a microtitre plate mixer.
4. Read the relative light units with a suitable luminometer (λ_{max} = 425 nm). Alternatively the substrate may be used in a luminometer which reads the signal in a test tube. For test tube applications increase substrate volumes as required.

References

1. Bergmeyer, H.-U. (1963) In *Methods of enzymatic analysis*, 2nd edn, p. 783. Academic Press, New York.
2. Moessner, E., Boll, M. and Pfleiderer, G. (1980) *Hoppe-Seyler's Z. Physiol. Chem.* **361**, 543.

3. Kresze, G. B. (1983) In: *Methods of enzymatic analysis*, 3rd edn. (ed. H. U. Bergmeyer and M. Grabor), Vol. 2, pp. 86–8. Verlag Chemie, Weinheim.
4. Doumas, B. T., Bayse, D. D., Carter, R. J., Peters, T. Jr. and Schaffer, R., (1981) *Clin Chem*. **27**, 1642–50 and 1651–54.
5. Thorne, C. J. R. (1978), *Techniques in Protein and Enzyme Biochemistry*, Pt. 1, **B104**, 1–18.
6. Welinder, K. G. (1979) *Eur. J. Biochem*. **96**, 483.
7. Bergmeyer, H.-U. (1974) In *Methods of enzymatic analysis*, Vol 1–4. Academic Press, New York.
8. Makinen, K. K. and Tenovuo, J. (1982) *Anal. Biochem*. **126**, 100.
9. Marate, A., Salvaye, R., Negre, A. and Dauste Blazy, L. (1983) *Eur. J. Biochem*. **133(2)**, June 15.
10. Hildebrandt, A. G. and Roots, I. (1975) *Arch.Biochem. Biophys*. **171**, 385.
11. Hermanson, G. T. (1996) *Bioconjugate Techniques*. Academic Press Inc., San Diego and London.
12. Nakane, P. K. and Kawaoi, A. (1974) *J. Histochem. Cytochem*. **22**, 1084.
13. Avrameas, S. and Ternynck, T. (1971) *Immunochemistry* **8**, 1175.
14. Yoshitake, S. (1983) *J. Biochem*. **92**, 1413.
15. Ellman, G. L. (1959), *Arch. Biochem. Biophys*. **82**, 70–7.
16. Duncan, R. J. S., Weston, P. D. and Wrigglesworth, R. (1983) *Anal. Biochem*. **132**, 68, 82, 70–7.
17. Goding, J. W. (1978), *J. Immunol. Meth*. **20**, 241–53.
18. Hames, B. D. and Rickwood. (1981) *Gel Electrophoresis of Proteins: A Practical Approach*. IRL Press, Oxford.
19. Josephy, P., Eling, T. and Mason, R. (1982), *J. Biol. Chem*. **257**, 7, 3669–75.
20. Gallati, H. (1985) *J. Clin. Chem. Clin. Biochem*. **15, (6),** 323–528.
21. Yamamoto, R., Kimura, S., Hattori, S., Ishiguro, Y. and Kato, K. (1983) *Clin Chem*, **29(1)**, 151–53. Jan. 1983.
22. Schaap, A. P., Sandison, M. D. and Handley, R. S. (1987) *Tetrahedron Lett*. **28**, 1159–62.
23. Moessner, E., Boll, M. and Pfleiderer, G. (1980) *Hoppe-Seyler's Z. Physiol. Chem*. **361**, 539–49.
24. McComb, R. B. and Bowers, G. N. (eds) (1979) *Alkaline phosphatase*, p 175. Plenum Press, New York and London.
25. Laemmli, U. K. (1970) *Nature* **227**, 680–5.
26. Sobocinski, S. L. (1975) *Anal. Biochem*. **64**, 284–8.
27. Ashwell, G. (1976) *Methods in enzymology*, Vol. VIII 6, pp. 93–4. Academic Press, New York and London.
28. Dubois, R. N., Gilles, A., Hamilton, R. G., Rebers, P. A. and Smith, J. (1956) *Anal. Chem*. **28**, 30.
29. Wilson, M. B. and Nakane, P. K. (1978) *Immunofluorescence and related staining techniques* (ed. W. Knapp, K. Holubar and G. Wick). Elsevier/North Holland Biomedical Press, Amsterdam–New York.
30. Righetti, P. G. and Cravaggi, T. (1976) *J. Chromatog*. **127**, 1–28.
31. Stadtman, T. C. (1961) In *The enzymes*, Vol. 5 (ed. P. D. Boyer, H. A. Lardy and M. Myrback), p. 55. Academic Press, New York.
32. Cox, R. P., Gilbert, P. and Griffin, M. J. (1967) *Biochem. J*. **105**, 155.
33. Morton, R. K. (1955) *Biochem. J*. **61**, 232, 240.

34. Fernley, H. N. and Walker, P. G. (1967) *Biochem. J.* **104**, 1011.
35. Fawaz, E. N. and Tejirian, A. (1972) *Hoppe—Seyler's Z. Physiol. Chem.* **353**, 1779.
36. Engvall, E. and Perlmann, P. (1971) *Immunochemistry* **8**, 871.
37. Kearney, J. F., Radbruch, A., Liesegang, B. and Rajewsky, K. (1979) *Immunology* **123**, 1548.
38. Schaap, A. P., Akhavan, H. and Romano, L. J. (1989) *Clin. Chem.* **35**, No 9, 1863–4.
39. Urdea, M. S., Kolberg, J., Clyne, J., Running, J. A., Besemer, D., Warner, B. and Sanchez-Pescader, R. (1989) *Clin. Chem.* **35**, No 8, 1571–5.
40. Jablonski, E., Moomaw, E. W., Tullis, R. H. and Ruth, J. L. (1986) *Nucleic Acid Res.* **14**, 6115–28.
41. O'Sullivan, M. J., Bridges, J. W. and Marks, V. (1979) *Ann. Clin. Biochem.* **16**, 221.
42. Williams, D. G. (1984) *J. Immunological Methods* **72**, 261–8.
43. Shannon, M. L., Kay, E. and Ley, J. Y. (1966) *Journal of Biochemical Chemistry* **241**, 2166–72.
44. Morita, Y., Yamashita, H., Mori, E., Kato, M. and Aibora, S. (1982) *J. Biochem.* **92**, 531–39.
45. Theorell, H. (1950) *Acta. Chem. Scand.* **4**, 22.
46. Imagwa, M., Yoshitake, S., Hamguchi, Y., Ishikawa, E., Niitsu, Y. and Urushizaki, I. (1982) *J. Appl. Biochem.* **4**, 41–57.
47. Hashida, S., Ishikawa, E. and Imagwa, W. (1983) *Anal. Lett.* **16** (B19), 1509–23.
48. Weilinder, K. G. (1979) *Eur. J. Biochem.* **96**, 483—502.
49. Cecil, R. and Ogston, A. G. (1951) *Biochem. J.* **49**, 105.
50. Paul, K. G. (1963) *The enzymes*, Vol. 8 (ed. P. D. Boyer). Academic Press, New York.
51. Barman, T. E. (1969) *Enzyme handbook*, Vol. 1, p. 234. Springer-Verlag, Berlin.
52. Habeeb, A. F. S. A. (1972) *Meth. Enzymol.* **25**, 457.
53. Fowler, A. V. and Zabin, I. (1978) *J. Biol. Chem.* **253**, 5521–5.
54. Craven, G. R., Steers, E. and Anfinsen, C. B. (1965) *J. Biol. Chem.* **240**, 2468–78.
55. Weber, K., Sund, H. and Wallenfeis, (1964) *Biochem. Z.* **339**, 498.
56. Bahl, O. P. and Agarwal, K. M. (1969) *J. Biol. Chem.* **244**, 2970.
57. Chandler, H. M., Cox, J. L., Healey, K., MacGregor, A., Premier, R. R. and Hurrell, J. G. (1982) *J. Immunol. Meth.* **53**, 187.
58. Meyerhoff, M. and Rechnitz, G. (1980) *Meth. Enzymol.* **70**, 439.
59. Avramacs, S. and Ternynck, T. (1971) *Immunochem.* **8**, 1175.
60. Canfield, R. E., Collins, J. C. and Sobel, J. H. (1974) In *Lysozyme* (ed. E. F. Osserman, R. E. Canfield and S. Beychock), p. 63. Academic Press, New York.
61. Banaszak, L. J. and Bradshaw, R. A. (1975) In *The enzymes*, Vol. 11, (ed. P. D. Boyer), p. 369. Academic Press, New York.
62. Rowley, G. L., Rubenstein, K. E., Huisjen, J. and Ullman, E. F. (1975) *J. Biol. Chem.* **250**, 3759.
63. Milhausen, M. and Levy, H. R. (1975) *Biochemistry* **5**, 453.
64. Olive, C. and Levy, H. R. (1971) *J. Biol. Chem.* **245**, 2043.
65. Tiggerman, R., Plattner, H., Rasched, I., Baeuerle, P. and Wachter, E. (1981) *J. Histochem. Cytochem.* **29**, 1387.
66. Uto, I., Ishimatsu, T., Hirayama, H., Veda, S., Tsuruta, J. and Kambara, T. (1991) *J. Immunol. Meth.* **138**, 87–94.

67. Bieniarz, C., Husain, M., Barnes, G., King, C. A. and Welch, C. J. (1996) *Bioconjugate Chem.* 88–95.
68. Kitagawa, T. and Aikawa, T. (1976) *J. Biochem* (Tokyo) **79**, 233–36.
69. Myers, D. E. (1989) *J. Immunol. Meth.* **121**, 129–42.
70. Iwai, K., Fukuoka, S., Fushiki, T., Kado, K., Sengoku, Y. and Semba, T. (1988) *Anal. Biochem.* **171**, 277–82.
71. Jue, R., Lambert, J. M., Pierce, L. R. and Trout, R. R. (1978) *Biochemistry* **17**, 5399.
72. Carlsson, J., Drevin, H. and Axen, R. (1978) *Biochem. J.* **173**, 723–37.
73. Ghetie, V., Till, M. A., Ghetie, M. A., Tucker, T., Porter, J. and Patzer, E. J. (1990) *Bioconjugate Chem.* **1**, 24–31.
74. Cumber, A. J., Forrester, J. A., Foxwell, B. M., Ross, W. C. and Thorpe, P. E. (1985) *Meth. Enzymol.* **112**, 207–25. Academic Press, New York.
75. Hermanson, G. T. (1996) *Bioconjugate techniques*, pp 542, 553, 568. San Diego Academic Press.
76. Joiris, E., Basin, B. and Thornback, J. A. (1991) *Nucl. Med. Biol.* **18**, 353–56.
77. McCracken, S. and Meighen, E. A. (1981) *J. Biol. Chem.* **256**, 3945.
78. Caswell, M. and Caplow, M. (1980) *Biochemistry* **19**, 2907.
79. Saunders, B. C., Holmes Siedle, A. G. and Stark, B. P. (1964) In *Peroxidase*, p. 271. Butterworths, London.
80. Marklund, S. (1973) *Arch. Biochem. Biophys.* **154**, 614.
81. Fowler, A. V. and Zabin, I. (1977) *Proc. Natl. Acad. Sci. USA* **74**, 1507.
82. Rothman, F. and Byrne, R. (1963) *J. Mol. Biol.* **6**, 330.
83. Bentley, R. (1963) In *The enzymes*, Vol. 7 (ed. P. D. Boyer, H. L. Lardy and H. Myrback), p. 567. Academic Press, New York.
84. Fowler, A. V. and Zabin, I. (1977) *Proc. Natl. Acad. Sci. USA* **74**, 1507.
85. Peterson, J., Harmon, K. M. and Niemann, C. (1948) *J. Biol. Chem.* **176**, 1.
86. Olive, C. and Levy, H. R. (1971) *J. Biol. Chem.* **245**, 2043.
87. Bos, E. S., Van der Doelen, A. A., Van Rooy, N. and Schuurs, A. H. (1981) *J. Immunoassay* **2** ,187–204.
88. Bovaird, J. H., Ngo, T. T. and Lewhoff, H. M. (1982) *J. Clin. Chem.* **28(12)**, 2423–26.
89. Avraneas, S. and Glibert, B. (1976) *Biochemie* **54**, 837.
90. Gallati, H. (1979) *J. Clin. Chem. Clin. Biochem.* **17**, 1.
91. Ellens, D. J. and Geilkens, A. J. (1980) *J. Immunol. Meth.* **37**, 325.
92. Ochoa, E. A. (1968) *Clin. Biochem. 2* **(I)**, 71–80.
93. Halpern, E. P., Rosoff, S. J. and Weiner, S. (1972) *Clin. Chem.* **18**, 593.
94. Pluzek, K. J. and Ramlau, J. (1988) *Handbook of Immunoblotting of Proteins,* **I**, (Bjerrum, O. J. and Heegaard, N. H. H. eds). CRC Press Inc., Boca Raton Fl 177.
95. Chandler, H. M., Cox, J. L., Healy, K., MacGregor, A., Premier, R. R. and Hurrell, J. G. (1982) *J. Immunol. Meth.* **53**,187.
96. Tanimori, H., Ishikawa, F. and Kitigawa, T. (1983) *J. Immunol. Meth.* **62**, 123.
97. Blake, M. S., Johnson, K. H., Russell Jones, J. G. and Gotochlich, E. C. (1984) *Anal. Biochem.* **136**, 175.
98. Larsson, L. I. (1988) *Immunol. and Chemistry: Theory and Practice*, 95 (INP).
99. Barlie, F. and Trombetta, L. D. (1982) *J. Histochem.* **5**, 12 (DAB).

100. Nakane, P. K. (1968) *J. Histochem Cytochem.* **16**, 557 (4CIN).
101. Graham, R. C. (1965) *J. Histochem Cytochem.* **13**, 150 (AEC).
102. Maniatis, T., Fritsch, E. and Sambrook, J. (1982) *Molecular closing: A laboratory manual.* Cold Spring Harbor Laboratory (BCIG).
103. Craven, G. R., Steers, J. T. E. and Anfinsen, C. B. (1965) *J. Biol Chem.* **240**, 2468–77.
104. Cornish, C. J. (1970) *Am. J. Clin. Path.* **53**, 68.
105. JBL Scientific Inc. Attophos^{TM}, Data Sheet Catalog No. 1660A.
106. Tuuminen, T., Pacomaki, P., Rakkolainen, A., Welin, M. G., Weber, T. and Kapyaho, K. (1991) *J. Immunoassay* **12,** 29–46.
107. Pierce Quenta Blue. TM Data sheet Product Number 15169.
108. Hofmann, J. and Sernetz, M. (1984) *Anal Chimica Acta* **163,** 67.
109. Ashihara, Y., Saruta, H., Ando, S., Kituchi, Y. and Kasahara., Y. (1994) In *Bioluminescence and Chemiluminescence: Fundamental and Applied Aspects*, 321–24. John Wiley & Sons, New York.
110. Bronstein, I., Voyta, J. C., Thorpe, G. H. G., Kricka, L. J. and Armstrong, G. (1989) *Clin. Chem.* **35,** 1441–6.
111. Fimbel, S., Dechaud, H., Grenot, C., Tabard, L., Claustrat, F., Bador, R. and Pugeat, M. (1995) *Steroids* **60,** 686–92.
112. Jordan, T., Walus, L., Velickovic, A., Last, T., Doctrow, S. and Liu, H. (1996) *J. Pharm. Biom. Anal.* **14,** 1653–62.
113. Legris, F., Martel-Pelletier, J., Pelletier, J. P., Colman, R. and Adam, A. (1994) *J. Immunol. Methods* **168,** 111–21.
114. Lehel, C., Daniel-Issakani, S., Brasseur, M. andStrulovici, B. (1997) *Anal. Biochem.* **244,** 340–6.
115. Nishizono, I., Iida, S., Suzuki, N., Kawada, H., Murakami, H., Ashihara, Y. and Okada, M. (1991) *Clin. Chem.* **37,** 1639–44.
116. Olesen, C. E. M., Voyta, J. C. and Bronstein, I. (1997) *Methods in Molecular Biology. Recombinant Proteins: Detection and Isolation Protocols* **63**, 71–6.
117. Roffman, E. and Frenkel, N. (1991) *J. Immunol. Methods* **138,** 129–31.
118. Thorpe, G. H. G., Bronstein, I., Kricka, L. J., Edwards, B. and Voyta, J. C. (1989) *Clin. Chem.* **35**, 2319–21.
119. Lumino Phos^{TM}, Lumigen Inc.
120. Seitz, W. R. (1984) *Clin. Biochem.* **17,** 120.
121. Whitehead, T. P., Thorpe, G. H. G., Carter, T. J. N, Groucutt, C. and Kricka, L. J. (1983) *Nature* **305**, 158.
122. Lespie, E., Moseley, S., Amess, R., Baggett, N. and Whitehead, T. P. (1985) *Anal. Biochem.* **145**, 96.
123. Kricka, L. J. and Thorpe, G. H. G. (1983) *Analyst* **108**, 1274–96.
124. Kricka, L. J. (1985) *Ligand-binder assays.* Marcel Dekker, New York.
125. Weeks, I., Campbell, A. K. and Woodhead, J. S. (1983) *Clin. Chem.* **29,** 1480–3.
126. Arakawa, H., Maeda, M., Tsuji, A. and Kambegawa, A. (1981) *Steroids* **28,** 453–64.
127. Carter, T. J. N., Groucutt, C. J., Stott, R. A. W., Thorpe, G. H. G and Whitehead, T. P. (1982) Enhanced luminescent or luminometric assay. *European Patent Publication* 87 959.
128. Thorpe, G. H. G., Kricka, L. J., Gillespie, E., Moseley, S., Amess, R., Baggett, N. and Whitehead, T. P. (1985) *Anal. Biochem.* **145,** 96–100.

129. Kricka, L. J., Thorpe, G. H. G and Whitehead, T. P. (1983) Enhanced luminescent or lumionometric assay. *European Patent Publication,* 116454.
130. Thorpe, G. H. G., Kricka, L. J., Moseley, S. B. and Whitehead, T. P. (1985) *Clin. Chem.* **31,** 1335–41.
131. Thorpe, G. H. G., Whitehead, T. P., Penn, R. and Kricka, L. J. (1984) *Clin. Chem.* **30,** 806–7.
132. Thorpe, G. H. G., Haggart, R., Kricka, L. J. and Whitehead, T. P. (1984) *Biochem. Biophys. Res. Commun.* **119,** 481–7.
133. Wang, H. X., Stott, R. A., Thorpe, G. H. G., Kricka, L. J., Holder, G. and Rudd, B. T. (1985) *Steriods* **44,** 317–28.
134. Pierce SuperSignal[R] LBA Data sheet product number 37070.

4

Time-resolved fluorescence immunoassay

GEOFF BARNARD

1. Introduction

Today, immunoassay methodology is applied in many branches and disciplines of scientific investigation. The procedures are essentially of two types, competitive and non-competitive, and historically these have been typified by radioimmunoassay (RIA) and immunoradiometric assay (IRMA), respectively (see Chapter 1). In recent years, however, there has been the widespread adoption of procedures using alternative detection systems that avoid the use of radioisotopes (e.g. enzymes, luminescent compounds etc.). In particular, over the last 10 years, it has become well established that timc-resolved fluorescence immunoassay (FIA) can offer unique advantages over most other non-radioisotopic procedures (1).

2. Principles of time-resolved fluorescence

Fluorescent molecules absorb light of a particular wavelength which excites their electronic fields from a resting to a higher energetic state. The return to the resting condition is accompanied by a non-luminescent conversion (e.g. heat) and, more particularly, by a luminescent transition directly to ground state (true fluorescence) or through a semi-stable triplet state (more commonly termed phosphorescence). The small decrease in luminescent energy which results in the difference between the excitation and emission wavelengths is defined as the Stokes' shift. In fluorescence this shift is normally 30 to 50 nm. The Stokes' shift of certain β-diketone chelates of several different lanthanides (e.g. europium or samarium), however, is considerably greater (see *Figure 1*) and is a factor that helps to minimize the background signal.

A second characteristic of fluorescence is its lifetime which can be defined as the decay rate of the excited state. The fluorescence from normal com-

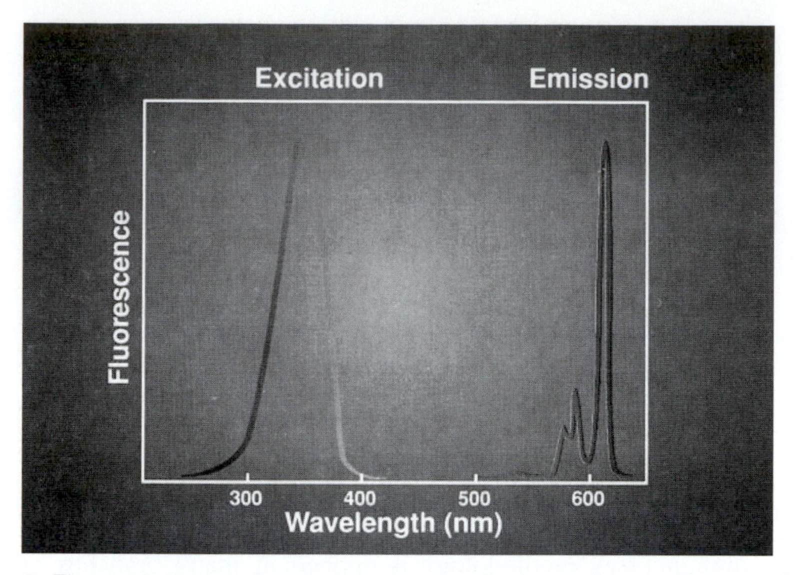

Figure 1. The excitation and emission spectra of a europium β-diketone chelate demonstrating a large Stokes' shift (used by permission from Wallac Oy, Turku, Finland).

ponents in assay systems (i.e. serum fluors and proteins, plastics, assay buffers etc.) is usually relatively short-lived, of nanosecond duration. On the other hand, the fluorescent lifetimes of the lanthanide chelates is up to six orders of magnitude longer (approximately 10^{-3} to 10^{-6} s). Consequently, the millisecond emission from these metal chelates can be distinguished easily from the background nanosecond fluorescence by using a gated fluorometer with time-resolution capability and employing appropriate delay, counting and cycle times (2). The measurement principle is illustrated in *Figure 2*.

Europium (Eu^{3+}) and samarium (Sm^{3+}) are two of several elements which comprise the lanthanide series. Europium has an atomic weight of 152 and is similar in size to the radioactive isotope iodine-125 which is commonly used in RIA and IRMA. Europium is a stable non-radioactive isotope and does not denature the material to which it is attached by radiolysis. However, europium requires a carrier molecule to form a conjugate with any immunoreactive component. The commercial availability of stable europium or samarium chelates of the EDTA derivative, DTTA [N^1-(*p*-isothiocyanatobenzyl)-diethylenetriamine N^1,N^2,N^3,N^3-tetraacetic acid], enables the research worker to produce labelled antibodies with high specific activity by very simple and reliable procedures (see *Protocols 1–3*).

The decay time and intensity of lanthanide fluorescence is dependent on the structure of the ligands which chelate the metal and on the nature of its physical environment (3). For example, the DTTA chelate has minimal fluorescence and in order to measure the fluorescence of europium or samar-

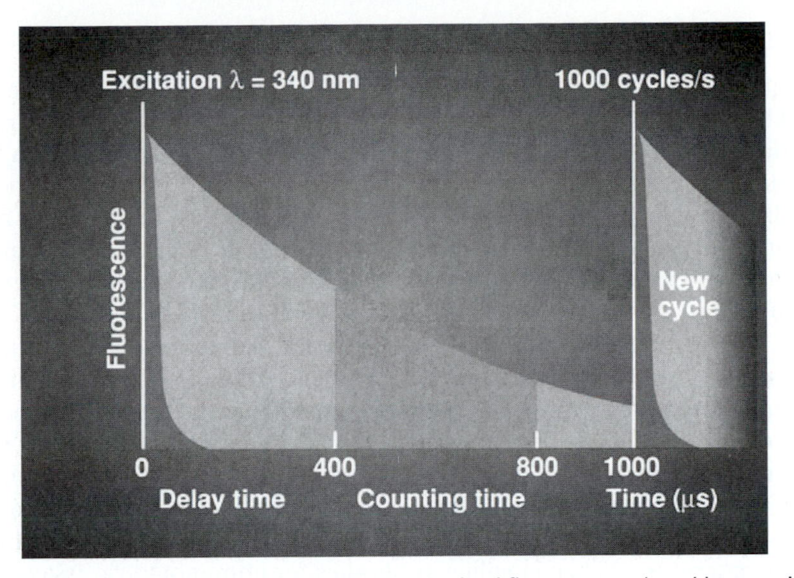

Figure 2. The measurement principle of time-resolved fluorescence (used by permission from Wallac Oy, Turku, Finland).

ium with high sensitivity, it is necessary to dissociate the element from the immunoreactive components. This is achieved simply by lowering the pH to 2–3 through the addition of an appropriate buffer (the 'enhancement solution') containing a β-diketone (e.g. 2-naphthoyl trifluoroacetate), which chelates the lanthanide at low pH and possesses the required properties to produce a chelate with a high intensity fluorescence.

In addition, it is necessary to expel water from the complex and this is accomplished by the presence, in the enhancement solution, of a detergent (Triton X-100), which together with a fatty acid derivative (trioctylphosphine oxide), dissolves the chelate and keeps it inside a micelle. The total immunoassay process may be summed up by the acronym DELFIA (Dissociation Enhanced Lanthanide FluoroImmunoAssay).

Fluorescence is initiated in the time-resolved fluorometer with a pulse of excitation energy, repeatedly and reproducibly. The complete cycle time for the measurement is typically one millisecond (i.e. delay phase—400 μs; measurement phase—400 μs; lag phase—200 μs). Thus, in one second, the fluorescent material can be pulse excited 1000 times with the concomitant accumulation of the generated signal. In addition, the wide dynamic range of the fluorometer enables the development of immunoassays that are not only sensitive but cover a working range of analyte concentration which can be up to five orders of magnitude (e.g. human chorionic gonadotrophin—analyte working range: 0.1 to 10000 IU/litre; instrument dynamic working range: 500 to 5000000 counts per second).

3. Protein labelling with lanthanide chelates

The labelling reagents (Eu^{3+} and Sm^{3+}) that are commercially available consist of bifunctional chelating agents with an isothiocyanate group which reacts with the ϵ-amino groups of the available lysine residues of the protein to form a stable, covalent thiourea bond. The coupling reaction is very mild, and the overall hydrophilic nature and negative net charge of the chelate allow labelling to a high specific activity without decreased affinity and immunoreactivity or increased non-specific binding. In addition, the thermodynamic stability of the chelate allows long-term storage of the labelled proteins.

The kinetics of the conjugation reaction depend upon the concentration of the reactants, pH and temperature. Typically, labelling is performed with an overnight incubation at pH 8.5 and at room temperature. Higher labelling yields can be obtained by extending the labelling reaction to 2–3 days at room temperature, or by the use of higher temperatures (up to 37°C). Furthermore, higher incorporation rates may be achieved by using a higher pH (up to 9.8). Alternatively, lower labelling yield can be achieved by using a lower pH or lower temperature. When labelling antibodies for immunometric assays, 5–15 Eu^{3+}/IgG may be optimal to yield an assay with high sensitivity and low background (typically <1000 counts per second). For many assays, lower labelling yield (i.e. <5 Eu^{3+}/IgG) may give perfectly acceptable results. Labelling antibodies to give a specific activity in excess of 20 Eu^{3+}/IgG may occasionally give an elevated background.

Protocol 1. Labelling antibodies with europium (or samarium) chelates

Equipment and reagents

- Fraction collector, peristaltic pump, UV-detector (e.g. Uvicord)
- Spectrophotometer for measurement at 280 nm
- Sephadex PD-10 disposable column (Pharmacia)
- Europium-labelling kit (Wallac 1244–302; 0.2 mg) or samarium-labelling kit (Wallac 1244–303; 0.2 mg)
- Labelling buffer:[a] Typically 50 mmol/litre $NaHCO_3$ (pH 8.5) containing 0.9% sodium chloride

Method

1. Pre-equilibrate Sephadex PD-10 column with labelling buffer. Check for a stable baseline by monitoring effluent using a UV-detector.
2. Load approximately 1 mg of antibody onto the column in the minimum of buffer solution. Wash the antibody through the column using the labelling buffer.
3. Monitor effluent using the UV detector and collect the eluent fractions containing the protein.

4. Check the protein concentration at 280 nm using the spectrophotometer and by applying the formulae: UV absorption of 1.34 = IgG concentration of 1 mg/ml.[b]
5. Dissolve the labelling reagent directly with typically 0.5 ml of the pre-treated protein in the labelling buffer.[c]
6. Gently mix the protein and the reagent and incubate overnight at room temperature.

[a]Buffers which contain free primary amines (e.g. Tris or glycine), secondary amines (e.g. Hepes, Mops, Bicine, etc.) and bacteriostatic agents (e.g. sodium azide) cannot be used.
[b]When small amounts or diluted solutions of proteins are to be labelled, it may be necessary to concentrate the sample by an appropriate method. Alternatively, buffer exchange may be carried out using dialysis.
[c]An appropriate amount of protein that can be labelled with 0.2 mg lanthanide chelate reagent is typically 1 mg.

Protocol 2. The purification of labelled proteins using disposable Sephadex PD-10 columns

Equipment and reagents

- Fraction collector, peristaltic pump, UV-detector (e.g. Uvicord)
- Spectrophotometer for measurement of protein at 280 nm
- Sephadex PD-10 disposable column (Pharmacia)
- Elution buffer: typically 50 mmol/litre Tris–HCl (pH 7.75) containing 0.9% sodium chloride and 0.05% sodium azide

Method

1. Pre-equilibrate Sephadex PD-10 column with elution buffer. Check a stable baseline by monitoring effluent using a UV-detector.
2. Load contents of the labelling reaction vial onto the column and wash the labelled antibody through the column using the elution buffer.
3. Monitor the eluate using the UV detector and collect the fractions containing the labelled protein.

The commercially available europium or samarium chelates are appropriate for labelling proteins or hapten–protein conjugates with a molecular weight in excess of 5000. Accordingly, the simple gel filtration technique described in *Protocol 3* can be used for the separation of the labelled protein from the free unreacted chelate. Nevertheless, the risk of unsatisfactory separation of the protein from free chelates and from possible aggregates remains. Alternatively, Sephadex G-25 or G-50 (1.5 × 30 cm column) may be used to obtain a more satisfactory separation of the labelled protein from unreacted chelate. In addition, Sepharose 6B (e.g. 1.5 × 40 cm column) is recommended for the fractionation and purification of labelled antibodies from unreacted chelates and from aggregated proteins. Consequently, by combining Sepharose 6B with Sephadex G-5O (e.g. 10 cm of Sephadex on the top of the Sepharose column) the resolution between free chelates and labelled proteins can be further improved.

For smaller peptides, haptens or other amine-containing compounds, alternative specific purification procedures will need to be developed.

Protocol 3. The characterization of labelled proteins

Equipment and reagents

- Spectrophotometer for measurement of protein at 280 nm
- Time-resolved fluorometer (Wallac 1234)
- Eu-standard: 100 nmol/litre Eu^{3+} in 0.1 mol/litre acetic acid[a]
- Enhancement solution (Wallac 1244–104)[a,b]
- Alternatively Sm-standard: 1000 nmol/litre Sm^{3+} in 0.1 mol/litre acetic acid[b]
- Stabilizer: 7.5% bovine serum albumin (BSA), highly purified from heavy metal contaminants[a,b]

Method

1. Dilute a small aliquot (i.e. 10 μl) of each fraction containing the Eu^{3+} or Sm^{3+}-labelled protein in enhancement solution (range: 1:10000–1:100000 or 1:1000–1:10000, respectively).
2. Measure the fluorescence in microtitration wells (200 μl/well; in duplicate) using the time-resolved fluorometer.
3. Compare results to the fluorescence obtained from the 1 nmol/litre Eu-standard or the 10 nmol/litre Sm-standard (stock standards, diluted 1:100 in enhancement solution).
4. Calculate the protein concentration of each fraction, using the UV spectrophotometer, from the absorbance at 280 nm after subtracting the absorbance of the aromatic thiourea bonds formed (0.008 μmol/litre).[c]
5. For IgG the following equations can be used:

$$\text{Protein concentration} = \frac{A(280) - (0.008 \times \text{Ln}^{3+}\ (\mu\text{mol/litre}))}{1.34}$$

$$\text{i.e. IgG } (\mu\text{mol/litre}) = \frac{\text{IgG (mg/ml)} \times 1000000}{160000\ \text{(g/mol)}}$$

$$\text{Yield, La}^{3+}/\text{IgG} = \frac{\text{La}^{3+}\ (\mu\text{mol/litre})}{\text{IgG } (\mu\text{mol/litre})}$$

$$\text{Recovery (mg)} = \frac{100 \times \text{IgG (mg/ml)} \times \text{volume pooled (ml)}}{\text{IgG added (mg)}}$$

Where $A(280)$ = measured absorbance of the pooled protein at 280 nm. Ln^{3+} = micromolar concentration of Eu^{3+} or Sm^{3+} in the eluate. Molecular weight of IgG is 160000 and absorptivity value (for 1 mg/ml) is 1.34.

[a]These materials are contained in the Eu-labelling kit (Wallac 1244–302).
[b]These materials are contained in the Sm-labelling kit (Wallac 1244–303).
[c]This absorption correction is not valid if the labelling yield exceeds 20 Eu^{3+} or Sm^{3+}/IgG or if the protein absorptivity value is >2 (for 1 mg/ml).

4. Storage of labelled proteins

Proteins labelled with the lanthanide chelates are usually very stable, which gives the labelled compounds a long shelf-life. Long-term work with the protein not only depends on the stability of the bound chelate but as much on the inherent stability of the particular protein.

To increase the stability of labelled antibodies, a stabilizer of highly purified BSA can be used as a carrier protein with a final concentration of 0.1%. The labelled proteins may be stored at 4°C. Repeated freezing and thawing should be avoided. In cases where BSA cannot be used as a carrier protein, the labelled proteins may be stored without the addition of extra protein provided the concentration of the labelled reagent is >50 μg/litre. If other proteins or compounds are used as carriers, they must be purified of any heavy metal contamination prior to addition. The carrier used must also be free of chelating agents.

The labelled proteins cannot be stored in phosphate buffer. In addition, the pH of the storage medium should be slightly alkaline (not below pH 7). If during storage of labelled reagent the background of the assay increases due to aggregation formation, the antibodies may be filtered through a 0.2 μm membrane.

5. Application to two-site immunometric assay of proteins

Many two-site IFMAs for the measurement of large molecular weight antigens (e.g. protein hormones) have been developed, evaluated and reported in international journals since the first publication in 1983 (4). The basis of these 'sandwich' assays is the direct or indirect immobilization of an excess of specific antibody to the polystyrene microtitre wells (i.e. the 'capture' antibody). During the primary antibody-binding reaction (one step) or following phase separation (two-step), a second europium-labelled antibody (i.e. the detecting antibody) with a different epitope specificity is added in excess.

After the immunoreaction has been completed the excess materials are washed away and, following the addition of enhancement solution, fluorescence is measured in the fluorometer. The signal is proportional to the concentration of the analyte.

There is an extensive range of commercially available two-site DELFIAs and these include: thyroid stimulating hormone (TSH); human chorionic gonadotrophin (hCG); α-fetoprotein (AFP); and human growth hormone among many others (E.G. & G. Wallac UK Ltd, Milton Keynes).

Protocol 4. A two-site time-resolved immunofluorometric assay for the measurement of a large molecular weight antigen using a europium-labelled detecting antibody

Equipment and reagents

- Time-resolved fluorometer (e.g. Wallac 1234)
- Plate washer (e.g. Wallac 1296–024)
- Plate shaker (e.g. Wallac 1296–001)
- Multichannel pipette, single channel pipettes, pipette tips
- Enhancement solution dispenser (e.g. Eppendorf Combipette, Socorex dispenser etc.)
- Two antibodies (preferably monoclonal) that have been well characterized and that recognize spatially separated epitopes on the surface of the antigen to be measured. One of these antibodies will have been labelled with europium according to *Protocols 1* and *2*
- NUNC Maxisorb microtitration plates
- Coating buffer: typically 0.1 M phosphate buffered saline (pH 7.4) containing 0.1% sodium azide
- Blocking buffer: typically 2% BSA in 0.1 M PBS (as above)
- Assay buffer: typically 0.05 M Tris–HCl (pH 7.75) containing 0.9% sodium chloride, 0.5% BSA, 0.1% sodium azide
- Wash solution: typically 0.01 M Tris–HCl (pH 7.8) containing 0.05% Tween 20 and 0.1% sodium azide
- Enhancement solution (Wallac 1244–101)

Method

1. Coat the required number of Nunc Maxisorp plates with the non-labelled monoclonal capture antibody at an appropriate dilution in coating buffer (2–5 μg/ml) by the addition of 200 μl per well using the multichannel pipette.
2. Cover plates with plastic adhesive sheet and incubate overnight at 4°C. Wash plate(s) three times using the plate washer. Tap plate(s) dry on absorbant paper.
3. Add 200 μl blocking buffer per well using the multichannel pipette. Recover with adhesive sheet. Incubate for at least 1 h at room temperature prior to use.[a] Wash plate(s) three times using the plate washer and tap plates dry on absorbant paper.
4. Add appropriate duplicate aliquots (typically 25 or 50 μl) of test sample or reference standard to designated wells.[b]

5. Add 200 μl of assay buffer to each microtitre well and incubate for 2 h at room temperature on the plate shaker. Wash plate(s) three times using the plate washer and tap plate(s) dry on absorbant paper.
6. Dilute europium-labelled monoclonal detecting antibody in assay buffer to an appropriate dilution (typically 0.5–5 μg/ml). Add 200 μl of diluted antibody solution to each microtitre well and incubate for 1 h at room temperature on the plate shaker. Wash plate(s) six times using the plate washer and tap plate(s) dry on absorbant paper.
7. Add 200 μl of enhancement solution to each microtitre well using the designated dispenser and incubate for 5 min at room temperature on the plate shaker.
8. Carefully transfer the plate to the time-resolved fluorometer and measure fluorescence according to the instrument manual.

[a]Sealed plates can be stored at 4 °C for several days prior to use.
[b]The range of the calibration curve should reflect the anticipated concentration range of the antigen present in the sample population.

Protocol 5. A two-site time-resolved immunofluorometric assay for the measurement of a large molecular weight antigen using a biotinylated detecting antibody and europium-labelled streptavidin

Equipment and reagents

- Time-resolved fluorometer (e.g. Wallac 1234)
- Plate washer (e.g. Wallac 1296–024)
- Plate shaker (e.g. Wallac 1296–001)
- Multichannel pipette, single channel pipettes, pipette tips
- Enhancement solution dispenser (e.g. Eppendorf Combipette, Socorex dispenser etc.)
- NUNC Maxisorb microtitration plates
- Two antibodies (preferably monoclonal) that have been well characterized and that recognize spatially separated epitopes on the surface of the antigen to be measured. One of these antibodies will have been biotinylated according to established procedures (5)
- Streptavidin labelled with europium according to *Protocols 1* and *2* or available from Wallac UK (Wallac 1244–360)
- Coating, blocking and assay buffers, wash and enhancement solutions as for *Protocol 4*

Method

1. Coat the required number of Nunc Maxisorp plates with the non-labelled monoclonal capture antibody at an appropriate dilution in coating buffer (2–5 μg/ml) by the addition of 200 μl using the multi-channel pipette.
2. Cover plates with plastic adhesive sheet and incubate overnight at 4 °C. Wash plate(s) three times using the plate washer. Tap plate(s) dry on absorbant paper.

Protocol 5. *Continued*

3. Add 200 μl blocking buffer per well using the multichannel pipette. Recover with adhesive sheet. Incubate for at least 1 h at room temperature prior to use.[a] Wash plate(s) three times using the plate washer and tap plate(s) dry on absorbant paper.
4. Add appropriate duplicate aliquots (typically 25 or 50 μl) of test sample or reference standard to designated wells.[b]
5. Add 200 μl of assay buffer to each microtitre well and incubate for 2 h at room temperature on the plate shaker. Wash plate(s) three times using the plate washer and tap plates dry on absorbant paper.
6. Dilute biotinylated monoclonal detecting antibody in assay buffer to an appropriate dilution (typically 0.5–5 μg/ml). Add 200 μl of diluted antibody solution to each microtitre well and incubate for 1 h at room temperature on the plate shaker. Wash plate(s) three times using the plate washer and tap dry on absorbant paper.
7. Dilute europium-labelled streptavidin in assay buffer to an appropriate dilution (typically 0.1–0.5 μg/ml). Add 200 μl of diluted streptavidin solution to each microtitre well and incubate for 30 min at room temperature on the plate shaker. Wash plate(s) six times using the plate washer and tap plates dry on absorbant paper.
8. Add 200 μl of enhancement solution to each microtitre well using the designated dispenser and incubate for 5 min at room temperature on the plate shaker.
9. Carefully transfer the plate to the time-resolved fluorometer and measure fluorescence according to the instrument manual.

[a]Sealed plates can be stored at 4°C for several days prior to use.
[b]The range of the calibration curve should reflect the anticipated concentration range of the antigen present in the sample population.

A published example of the above protocol is the development of a two-site immunofluorometric assay for the measurement of somatotropin in human serum (6). The streptavidin–biotin interaction provides significant amplification of the signal and results in a useful increase in overall assay sensitivity. In addition, europium–streptavidin may be used in combination with most biotinylated immunocomponents in place of enzyme-labelled streptavidin or avidin.

6. Application to competitive binding immunoassays of haptens

The first competitive binding FIAs for the measurement of small molecules in biological fluids were reported in 1985 (cortisol) and 1986 (digoxin) (1). In these systems, the analyte and a defined amount of solid-phase hapten–protein conjugate competed for a limited concentration of Eu-labelled antibodies.

Protocol 6. A competitive time-resolved fluoroimmunoassay for the measurement of a small molecular weight analyte using a solid-phase antigen and a europium-labelled monoclonal antibody

Equipment and reagents

- Time-resolved fluorometer (e.g. Wallac 1234)
- Plate washer (e.g. Wallac 1296–024)
- Plate shaker (e.g. Wallac 1296–001)
- Multichannel pipette, single channel pipettes, pipette tips
- Enhancement solution dispenser (e.g. Eppendorf Combipette, Socorex dispenser etc.)
- NUNC Maxisorb microtitration plates
- A monoclonal anti-hapten antibody that has been well characterized and labelled with europium according to *Protocols 1* and *2*
- A hapten–protein conjugate prepared using established procedures (7) using a protein that is not the same as that used in the synthesis of the immunogen
- Coating, blocking and assay buffers, wash and enhancement solutions as for *Protocol 4*

Method

1. Coat the required number of Nunc Maxisorp plates with the hapten–protein conjugate at an appropriate dilution in coating buffer by the addition of 200 μl using the multichannel pipette.[a]
2. Cover plates with plastic adhesive sheet and incubate overnight at 4°C. Subsequently, wash plate(s) three times using the plate washer. Tap plate(s) dry on absorbant paper.
3. Add 200 μl blocking buffer per well using a multichannel pipette. Recover with adhesive sheet. Incubate for at least 1 h at room temperature prior to use.[b] Wash plate(s) three times using the plate washer and tap plate(s) dry on absorbant paper.
4. Add appropriate duplicate aliquots (typically 25 or 50 μl) of test sample or reference standard to designated wells.[c]
5. Add 200 μl of assay buffer containing an appropriate dilution of europium-labelled anti-hapten antibody to each microtitre well, and incubate for 2 h at room temperature on the plate shaker.[a] Wash plate(s) six times using the plate washer and tap plates dry on absorbant paper.
6. Add 200 μl of enhancement solution to each microtitre well using the designated dispenser and incubate for 5 min at room temperature on the plate shaker.

Protocol 6. *Continued*

7. Carefully transfer the plate to the time-resolved fluorometer and measure fluorescence according to the instrument manual.

[a]The optimum concentration of the solid-phase antigen and europium-labelled anti-hapten antibody must be established for each assay. According to the principles of competitive immunoassay, the lower the concentration of competing analyte, the more sensitive the assay.
[b]Sealed plates can be stored at 4°C for several days prior to use.
[c]The range of the calibration curve should reflect the anticipated concentration range of the antigen present in the sample population.

An alternative approach to the development of competitive FIAs using solid-phase antigens is to use an unlabelled monoclonal or polyclonal antibody and detect its presence on the solid-phase by using europium-labelled Protein-G (Wallac 1244–361) or europium-labelled second antibody (e.g. anti-mouse or anti-rabbit).

7. Europium-labelled hapten assay

More recently, however, directly labelled europium–hapten conjugates have become available and the use of these reagents have formed the basis of the current commercial assays for the measurement of haptens by DELFIA (8). These include FIAs for the direct measurement of the three major sex steroid hormones in serum (oestradiol, progesterone and testosterone) as well as kits for the determination of serum total and free thyroxine (T4) and triiodothryronine (T3).

Protocol 7. A competitive time-resolved fluoroimmunoassay for the measurement of a small molecular weight analyte using a europium-labelled hapten

Equipment and reagents

- Time-resolved fluorometer (e.g. Wallac 1234)
- Plate washer (e.g. Wallac 1296–024)
- Plate shaker (e.g. Wallac 1296–001)
- Multichannel pipette, single channel pipettes, pipette tips
- Enhancement solution dispenser (e.g. Eppendorf Combipette, Socorex dispenser etc.)
- NUNC Maxisorb microtitration plates
- A polyclonal anti-hapten antibody that has been well characterized
- An appropriate anti-species second antibody (e.g. affinity purified anti-rabbit IgG)
- Europium-labelled hapten available from Wallac Oy, Turku, Finland
- Alternatively, a hapten–protein or polylysine conjugate can be prepared using established procedures (7) and labelled with europium using *Protocols 1* and *2*
- Coating, blocking and assay buffers, wash and enhancement solutions as for *Protocol 4*

Method

1. Coat the required number of Nunc Maxisorp plates with the anti-species second antibody (e.g. affinity purified anti-rabbit IgG) at an appropriate dilution (typically 2–5 μg/ml) in coating buffer by the addition of 200 μl using the multichannel pipette.
2. Cover plates with plastic adhesive sheet and incubate overnight at 4°C. Wash plate(s) three times using the plate washer. Tap plate(s) dry on absorbant paper.
3. Add 200 μl blocking buffer per well using the multichannel pipette. Re-cover with adhesive sheet. Incubate for at least 1 h at room temperature prior to use.[a] Wash plate(s) three times using the plate washer and tap plate(s) dry on absorbant paper.
4. Add appropriate duplicate aliquots (typically 25 or 50 μl) of test sample or reference standard to designated wells.[b]
5. Add 100 μl of assay buffer containing an appropriate dilution of europium-labelled hapten to each microtitre well.[c]
6. Add 100 μl of assay buffer containing an appropriate dilution of anti-hapten polyclonal antibody to each microtitre well.[c] Incubate for 2 h at room temperature on the plate shaker. Wash plate(s) six times using the plate washer and tap plate(s) dry on absorbant paper.
7. Add 200 μl of enhancement solution to each microtitre well using the designated dispenser and incubate for 5 min at room temperature on the plate shaker.
8. Carefully transfer the plate to the time-resolved fluorometer and measure fluorescence according to the instrument manual.

[a]Sealed plates can be stored at 4°C for several days prior to use.
[b]The range of the calibration curve should reflect the anticipated concentration range of the antigen present in the sample population.
[c]The optimum concentrations of the labelled hapten and specific antibody must be established for each assay. According to the principles of competitive immunoassay, the lower the concentrations of these reagents, the more sensitive the assay will become.

8. Monoclonal antibody screening procedures

Unquestionably, the use of time-resolved fluorescence has become my method of choice for monoclonal antibody screening. The microtitre plate format together with the sensitivity of detection has revolutionized antibody screening methods which can be very labour intensive.

Protocol 8. A generalized scheme for monoclonal antibody screening

Equipment and reagents

- Time-resolved fluorometer (e.g. Wallac 1234)
- Plate washer (e.g. Wallac 1296–024)
- Plate shaker (e.g. Wallac 1296–001)
- Multichannel pipette, single channel pipettes, pipette tips
- Enhancement solution dispenser (e.g. Eppendorf Combipette, Socorex dispenser etc.)
- NUNC Maxisorb microtitration plates
- The immunogen (if large molecular weight protein) or an appropriate hapten–protein conjugate prepared using established procedures (7), using a protein that is not the same as that used in the synthesis of the immunogen used in the production of monoclonal anti-hapten antibodies
- A detecting reagent (e.g. europium labelled anti-mouse IgG or europium-labelled Protein G) prepared using *Protocols 1* and *2* or commercially available from Wallac (Wallac 1244–130; 1244–361)
- Coating, blocking and assay buffers, wash and enhancement solutions as for *Protocol 4*

Method

1. Coat the required number of Nunc Maxisorp plates with the immunogen or hapten–protein conjugate at an appropriate dilution (2–5 μg/ml) in coating buffer by the addition of 200 μl using the multichannel pipette.
2. Cover plates with plastic adhesive sheet and incubate overnight at 4 °C. Wash plate(s) three times using the plate washer. Tap plate(s) dry on absorbant paper.
3. Add 200 μl blocking buffer per well using a multichannel pipette. Recover with adhesive sheet. Incubate for at least 1 h at room temperature prior to use.[a] Wash plates three times using the plate washer and tap plates dry on absorbant paper.
4. Add an appropriate aliquot (typically 25–200 μl) of culture supernatant to designated wells. If small aliquots (<50 μl) of culture supernatant are used, add a further 200 μl of assay buffer to each microtitre well and incubate for 2 h at room temperature on the plate shaker. Wash plate(s) three times using the plate washer and tap plate(s) dry on absorbant paper.
5. To each microtitre well, add 200 μl of detecting reagent (e.g. europium-labelled anti-mouse IgG or europium-labelled Protein G) suitably diluted in assay buffer (typically 0.5 μg/ml). Incubate for 1 h at room temperature on the plate shaker. Wash plate(s) six times using the plate washer and tap plates dry on absorbant paper.
6. Add 200 μl of enhancement solution to each microtitre well using the designated dispenser and incubate for 5 min at room temperature on the plate shaker.
7. Carefully transfer the plate to the time-resolved fluorometer and measure fluorescence according to the instrument manual.

[a]Sealed plates can be stored at 4 °C for several days prior to use.

9. Novel research assays

A clear example of the potential of this technology can be illustrated by our recent development of a novel non-competitive immunoassay for the measurement of small molecular weight antigens, which is termed idiometric assay (9). The use of time-resolved fluorescence facilitated the screening, identification, production and application of two types of monoclonal anti-idiotypic antibodies that recognize different epitopes in the variable region of a primary immunoglobulin molecule. These reagents were classified as:

(i) betatypes, which recognize the binding site of the antibody and compete with the antigen for binding; and
(ii) alphatypes, which recognize the framework region of the variable region of the antibody and are not sensitive to the presence or absence of the antigen.

The application of these two anti-idiotypes together with time-resolved fluorescence enabled the development of a non-competitive method for the measurement of oestradiol in serum (9).

10. Non-separation time-resolved fluoroimmunoassay

In 1989, we described (10) the use of a novel fluorescent oestrone-3-glucuronyl–europium derivative to develop a simple, rapid and precise homogeneous time-resolved FIA for the measurement of urinary oestrone-3-glucuronide (EG). The principle of the assay was based on the finding that the fluorescence from the lanthanide chelate was quenched when the labelled antigen was bound to a specific antibody present in solution. The method required no separation or enhancement solution since the novel chelate was fluorescent in the aqueous phase.

Protocol 9. Non-separation time-resolved fluoroimmunoassay to measure oestrone-3-glucuronide in samples of early morning urine to monitor ovarian function

Equipment and reagents

- Time-resolved fluorometer (e.g. Wallac 1234)
- Plate shaker (e.g. Wallac 1296–001)
- Multichannel pipette, single channel pipettes, pipette tips
- NUNC Maxisorb microtitration plates
- Polyclonal or monoclonal antibodies to oestrone-3-glucuronyl–BSA
- Assay buffer: Tris HCl buffer prepared by dissolving 6 g (50 mmol) of Tris in 1 litre of doubly distilled water containing 5 g of BSA, 0.5 g of bovine gamma globulin, 20 μmol diethylenetriamine pentaacetic acid, 0.5 ml of Tween 20, 9 g of NaCl, 0.5 g of NaN_3 and enough HCl to adjust the pH to pH 7.75
- Samples: Daily samples of early morning urine (EMU), collected from healthy, non-pregnant female volunteers throughout complete menstrual cycles and stored at −20 °C until analysis

Protocol 1. *Continued*

Method

A. Preparation of labelled antigen. This has been described in detail (10).

B. Standards:

Prepare oestrone-3-glucuronide (EG) standards in undiluted urine obtained from a male volunteer (EG concentration: approximately 10 nmol/litre) to cover the range of 10 to 458 nmol/litre.

C. Assay

1. Add 10 μl of standard or sample (undiluted urine) to microtitre wells in duplicate.
2. Add 100 μl of FIA buffer containing an appropriate concentration of polyclonal or monoclonal antibodies to EG-BSA, and 100 μl of FIA buffer containing 2 ng of EG–europium conjugate. Incubate the antibody–antigen binding reaction at room temperature for 10 min using the plate shaker.
3. Transfer the plate to the time-resolved fluorometer. Determine the unknown values by comparison with calibration curves (signal vs concentration of EG, nmol/litre).

In terms of specificity and accuracy, the characteristics of the rapid non-separation assay were comparable to both a separation FIA and RIA. In addition, the advantages of non-separation immunoassay include:

(i) simplicity;

(ii) speed;

(iii) increased precision; and

(iv) the potential for total automation.

Furthermore, the principle of time-resolution minimizes the problems that affect most other non-separation methods, which are subject to non-specific interference from factors in biological materials, assay buffers, reagents and plastics. Consequently, with the synthesis of novel lanthanide chelates with increased quantum yields of fluorescence in the aqueous phase, we can envisage a gradual switch over from separation to non-separation methodology for the fully automated measurement of the majority of common analytes by time-resolved fluorescence.

11. Simultaneous immunoassays

It has been shown that the fluorescence emission wavelengths from alternative lanthanides are sufficiently different to be able to discriminate between them using appropriate filters (see *Figure 3*). This observation has been

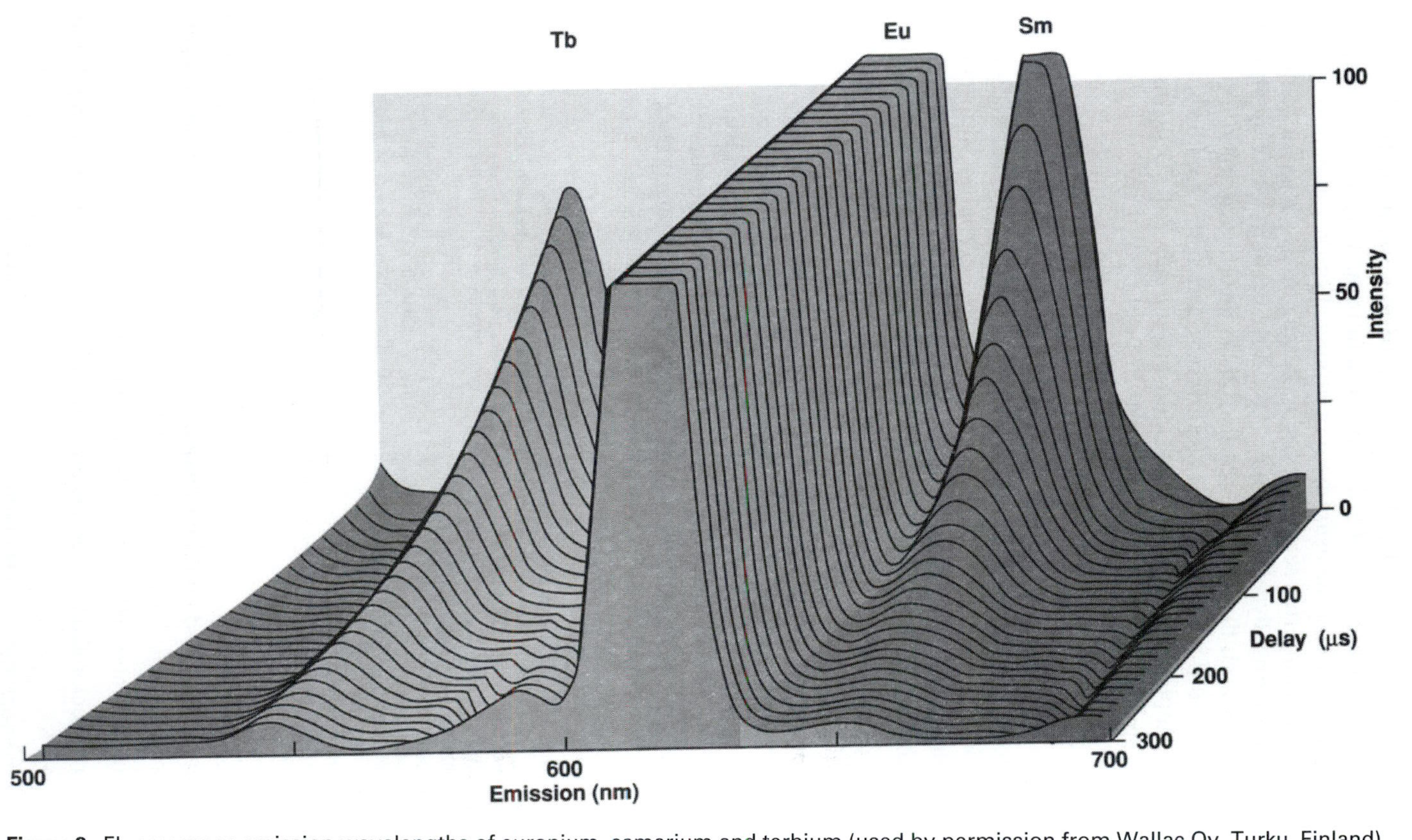

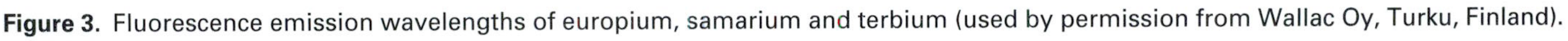
Figure 3. Fluorescence emission wavelengths of europium, samarium and terbium (used by permission from Wallac Oy, Turku, Finland).

successfully exploited in the development of a separation IFMA for the simultaneous measurement of follitropin and lutropin (11). In addition, a dual kit for the simultaneous measurement of human chorionic gonadotrophin and α-fetoprotein in maternal serum for the screening of Down's syndrome is now available commercially. To facilitate the development of dual and triple assays, the research fluorometer (Wallac 1234) has the capability of measuring up to three different lanthanides (europium, samarium, and terbium) with the use of simple emission filters. In addition, the new VICTOR instrument has the capability of measuring four different lanthanides (europium, samarium, terbium and dysprosium) together with prompt fluorescence, chemiluminescence and absorbance.

Most recently, we have developed a triple simultaneous FIA for the measurement of urinary oestrone-3-glucuronide (EG), pregnanediol-3-glucuronide (PG) and luteinizing hormone (LH) using europium, samarium and terbium chelates to investigate disorders of ovarian function. The assay involves the passive immobilization of three antibodies to the walls of polystyrene microtitre wells. After washing, 25 μl of undiluted urine or mixed standard are added to appropriate wells followed by the addition of 200 μl assay buffer containing all the labelled reactants.

Protocol 10. Simultaneous measurement of three urinary metabolites by time-resolved fluorescence immunoassay using europium, samarium and terbium labelled reagents

Equipment and reagents

- Time-resolved fluorometer (e.g. Wallac 1234)
- Plate washer (e.g. Wallac 1296–024)
- Plate shaker (e.g. Wallac 1296–001)
- Multichannel pipette, single channel pipettes, pipette tips
- Enhancement solution dispenser (e.g. Eppendorf Combipette, Socorex dispenser etc.)
- NUNC Maxisorb microtitration plates
- Monoclonal antibodies to PG–BSA, anti-EG–BSA and LH
- Europium-labelled PG, samarium-labelled anti-EG–BSA and terbium-labelled anti-LH
- Coating, blocking and assay buffers, wash and enhancement solutions as for *Protocol 4*
- Terbium enhancement additive (available from Wallac Oy, Turku, Finland)
- Samples: Daily samples of early morning urine (EMU), collected from female volunteers throughout complete menstrual cycles and stored at –20 °C until analysis

Method

1. Coat the required number of Nunc Maxisorp plates by adding, to each microtitre well, 200 μl of coating buffer containing:
   (a) monoclonal antibody to PG–BSA (1:1000 v/v);
   (b) monoclonal anti-idiotypic antibody to anti-EG–BSA antibody (1:100 v/v);
   (c) monoclonal anti-LH capture antibody (1:1000 v/v).

2. Cover plates with plastic adhesive sheet and incubate overnight at 4°C. Wash plate(s) three times using the plate washer. Tap plate(s) dry on absorbant paper.
3. Add 200 μl blocking buffer per well using the multichannel pipette. Re-cover with adhesive sheet. Incubate for at least 1 h at room temperature prior to use.[a] Wash plate(s) three times using the plate washer and tap plate(s) dry on absorbant paper.
4. To designated wells, add 25 μl of sample (undiluted EMU) or mixed standard solution containing:
   (a) EG (range 0–4480 nmol/litre);
   (b) PG (range 0–200 μmol/litre);
   (c) LH (range 0–250 IU/litre).
5. To each microtitre well, add 200 μl of assay buffer containing:
   (a) europium-labelled PG (1:1000 v/v);
   (b) samarium-labelled antibody to EG (1:200 v/v);
   (c) terbium-labelled antibody to LH (1:1000 v/v).
6. Incubate for 3 h at room temperature on the plate shaker. Wash plate(s) six times using the plate washer and tap plate(s) dry on absorbant paper.
7. Add 200 μl of enhancement solution to each microtitre well using the designated dispenser and incubate for 5 min at room temperature on the plate shaker.
8. Carefully transfer the plate to the time-resolved fluorometer and measure europium and samarium fluorescence according to the instrument manual.
9. Remove plate from the fluorometer and add 20 μl of terbium enhancement additive into each microtitre well. After shaking for 5 min on the plate shaker, return the plate to the fluorometer and measure terbium fluorescence.

[a]Sealed plates can be stored at 4°C for several days prior to use.

The method demonstrates appropriate sensitivity and excellent precision (all CVs $<10\%$) across the relevant working ranges for each hormonal parameter. The technique has been applied to serial EMUs collected from women with normal menstrual cycles (see *Figure 4*) and from those undergoing various therapies which include:

(1) ovarian stimulation with gonadotrophins;
(2) hormone replacement following the menopause; and
(3) treatment for pre-menstrual tension syndrome.

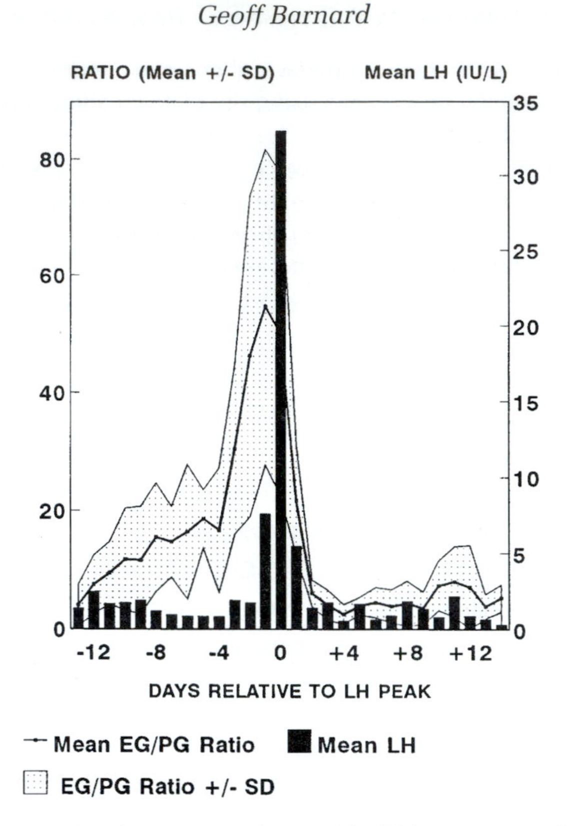

Figure 4. The mean ratio of oestrone-3-glucuronide (EG) to pregnanediol-3-glucuronide (PG) and the mean concentration of luteinizing hormone (LH) in samples of early morning urine collected throughout six normal menstrual cycles as determined by a simultaneous time-resolved fluorescence immunoassay of the three analytes.

The advantage to the patient is the ease of serial sample collection which can be performed at home. The advantage to the clinician is an overview of hormonal activity that cannot be obtained from the results obtained from serum samples taken at infrequent intervals.

12. Conclusion

Currently, we are developing FIAs and IFMAs for many other analytes for which there are no available commercial kits. The simplicity of the labelling procedure, the stability of the labelled reagents and the convenience of varied assay configurations has excellent potential in both research and routine environments.

Although outside the scope of this book, it may be noted that the advantages that time-resolved fluorescence has brought to immunoassay are now available to other laboratory procedures that have relied up to the present on the use of radioisotopes. For example, DNA hybridization (12) and the study of cytotoxicity (13) are being revolutionized by the adoption of lanthanide chelates as labels. In particular, the development of sandwich hybridization assays using microtitre plate format will facilitate the widespread introduction of DNA technology into the routine clinical laboratory.

In this chapter, I have concentrated on established DELFIA procedures pioneered by Wallac in Finland. My reasons for doing this have been my own experience over the last 15 years and to provide basic protocols that can be established, in principle and in practice, by the readers of this chapter in their own laboratories. Other time-resolved fluorescence systems utilizing lanthanide chelate chemistry have been and are being developed, most notably by Diamandis and his group (14) in Canada and by Mathis and his group in France (15).

Acknowledgements

I acknowledge with deep gratitude the contributions made in my own career by my dear friends Professor Bill Collins of the Diagnostics Research Unit, King's College School of Medicine, London, UK and Dr Fortune Kohen of the Department of Biological Regulation, The Weizmann Institute of Science, Rehovot, Israel. In addition, I am profoundly indebted to Wallac (both in Finland and in the UK) for their steadfast support over the last 20 years and for their generous provision of equipment, reagents and much encouragement.

Appendix: List of suppliers

Wallac Equipment and reagents:
- Wallac Oy, PO Box 10, FIN-20101 Turku, Finland
- E.G. & G. Wallac UK Ltd, 20 Vincent Avenue, Crownhill Business Centre, Crownhill, Milton Keynes, MK8 0AB

NUNC Maxisorp plates and plastic seal:
- Life Technologies Ltd, PO Box 35, Trident House, Renfrew Road, Paisley PA3 4EF

Eppendorf pipettes and other laboratory equipment and reagents:
- BDH/Merck Ltd., Hunter Boulevard, Magna Park, Lutterworth, Leicestershire

References

1. Barnard, G., Williams, J. L., Paton, A. C. and Shah, H. P. (1988). In *Complementary immunoassays* (ed. W. P. Collins), pp. 149–67. Wiley, Chichester.
2. Soini, E. and Kojola, H. (1983). *Clin. Chem.*, **29**, 65.
3. Hemmila, I., Dakubu, S., Mukkala, V-M., Siitari, H. and Lovgren, T. (1984). *Anal. Biochem.*, **137**, 335.
4. Hemmila, I. A. (1991) *Applications of fluorescence in immunoassays*. Wiley, Chichester.
5. Bayer, E. A. and Wilchek, M. (1980). *Methods Biochem. Anal.*, **26**, 1.
6. Strasburger, C., Barnard, G., Toldo, L., Zarmi, B., Zadik, Z., Kowarski, A. and Kohen, F. (1989). *Clin. Chem.*, **35**, 913.
7. Kohen, F., Bauminger, S. and Lindner, H. R. (1975). In *Steroid immunoassay* (ed. E. H. Cameron, S. G. Hillier, and K. Griffiths) pp. 11–32. Alpha Omega, Cardiff.
8. Mikola, H., Sundell, A-C. and Hanninen, E. (1993). *Steroids*, **58**, 330.
9. Barnard, G. and Kohen, F. (1990). *Clin. Chem.*, **36**, 1945.
10. Barnard, G., Kohen, F., Mikola, H. and Lovgren, T. (1989). *Clin. Chem.*, **35**, 555.
11. Hemmila, I., Holtinen, S., Pettersson, K. and Lovgren, T. (1987). *Clin. Chem.*, **33**, 2281.
12. Dahlen, P., Syvanen, A-C., Hurskainen, P., Kwiatkowski, M., Sund, C., Ylikoski, J., Soderlund, H. and Lovgren, T. (1987). *Mol. Cell. Probes*, **1**, 159.
13. Maley, D. T. and Simon, P. (1990). *J. Immunol. Meth.*, **134**, 61.
14. Diamandis, E. P. (1988). *Clin. Biochem.*, **21**, 139.
15. Mathis, G. (1993). *Annales Biologie Clinique*, **51**, abstract 27.

5

Light scattering techniques

DAVID J. NEWMAN, HANSA THAKKAR and CHRISTOPHER P. PRICE

1. Introduction

Quantitative immunoaggregation assays are very widely used due to their inherent simplicity and, in the liquid phase anyway, their ready automation, excellent precision and very fast reaction times. Gel-based immunoaggregation assays are certainly simple and robust but this chapter will focus on the various liquid phase techniques. Amongst the various homogeneous immunoassay technologies available, immunoaggregation assays are one of the few that are available to research scientists who wish to develop their own assays, as they can be carried out not only in sophisticated automated instruments but also in virtually any good quality spectrophotometer. They can be used for both haptens and proteins in most biological matrices and for analytes across the pM to mM concentration range; furthermore, they can use monoclonal and polyclonal antibodies both intact and fragment. Thus, using the appropriate detection system, immunoaggregation assays can compete very effectively, in terms of detection limits and convenience, with many enzyme and radioisotopically labelled immunoassays, and perform significantly better in terms of precision and ease of automation. The commonest techniques are immunoturbidimetry and immunonephelometry with and without the use of particle enhancement. These involve different aspects of light scattering; however, as turbidity can be measured on any spectrophotometer and nephelometry requires specialized equipment that may not be available in every laboratory, the main focus will be on the use of turbidimetric light scattering.

2. Fundamental aspects of immunoaggregation

There have been many reviews on various aspects of immunoaggregation assays a selection of which (1–5) are included in the reference section for background information. The principle of an immunoaggregation assay is the reaction between a polyvalent (multi-epitopic) antigen and a bivalent antibody that produces an immunoaggregate of sufficient size and frequency to increase the amount of light scattered when a beam of light passes through the reaction mixture. Ultimately this process can result in the formation of an immunoprecipitate hence the term immunoprecipitation, the theory of which was originally developed by Heidelberger and Kendall back in 1935 (6). The development of an immunocomplex involves several stages; the first and primary interaction involves van der Waals and Coulombic forces; this mainly electrostatic process causes the formation of small antigen:antibody complexes and is very rapid, occurring over a period of seconds to a few minutes. The secondary phase involves the stabilization of the small complexes and the generation of macromolecular complexes, which may precipitate. This phase is predominantly hydrophobic in nature but does involve some van der Waals and hydrogen bonding interactions; this secondary phase is essentially non-specific, i.e. it does not involve the antibody:antigen binding site, slower and essentially irreversible. The control of immunoaggregation reactions thus requires an appreciation of these intermolecular forces and how they can be manipulated.

An antibody:antigen complex can eventually reach between 50 and 100 nm in size, but over the reaction periods usually monitored, 2–10 minutes, is unlikely to exceed 20 nm. Size of the complex is important in selecting the optimum detection system as the relationship between complex size and incident wavelength determines the pattern of light scattering that occurs. It is not appropriate to go into the details of light scattering theory here, first developed by Lord Rayleigh in 1871 (7), but if the ratio between size of the complex and the incident wavelength reaches >1/10, then light is increasingly scattered in the forward direction and the refractive index of the complex has an increasing influence, the higher the refractive index the greater the scattering (*Figure 1*).

This can be demonstrated practically by taking particle standards of different sizes and looking at the absorbance at different wavelengths as shown in *Figure 2*. Taking a small particle (<40 nm) and a wavelength of about 340 nm, the maximum increase in scattering (absorbance) occurs when the particle size doubles to 80–100 nm. Although this is a model and not truly representative of an agglutination reaction, where simple doublets are rare and the aggregates are hardly simply spheres, this selection of wavelength has still proved successful for both non-enhanced and enhanced methodologies.

The agglutination reaction has a characteristic shaped calibration curve As the antigen concentration increases, a point is reached (see *Figure 3*) where there is an equivalence in the number of antigen molecules and the number of available antibody binding sites. If further antigen is added to the reaction, there is an excess of antigen to binding sites and the amount of immune complex decreases. This phenomenon of antigen excess can result in two antigen concentrations giving the same signal change and careful assay optimization is essential to ensure an appropriate working range.

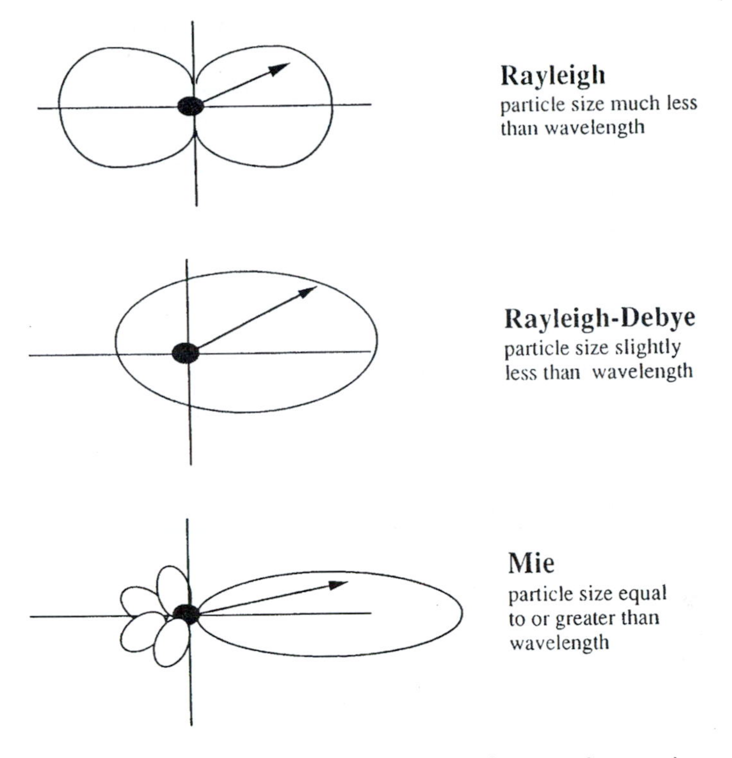

Figure 1. Light scattering theory. The figures represent the scattering envelope produced when light is shone on a single particle. The arrow represents the intensity of light scattered in one direction.

One way of avoiding antigen excess difficulties is to use an inhibition approach whereby the antigen is conjugated to a core, e.g. a latex particle, and the antibody concentration is set so as to be able to cause maximal agglutination but without reaching an excess. Free antigen can then inhibit the agglutination and cause a decrease in signal, a gross excess of antigen can only cause complete inhibition of agglutination. The use of an inhibition assay also enables the use of agglutination assays to be extended to haptens, as conjugation to a 'core' generates a 'multi-epitopic hapten'. The three types of assay format are shown in cartoon form in *Figure 4*, PETIA (Particle Enhanced Turbidimetric ImmunoAssay) which is analogous to the non-enhanced direct agglutination assay and PETINIA (Particle Enhanced Turbidimetric Inhibition Assay).

2.1 Monitoring techniques

There are two basic means of monitoring light scattering, that is turbidimetry and nephelometry (see *Figure 5*). Turbidimetry is the measurement of the

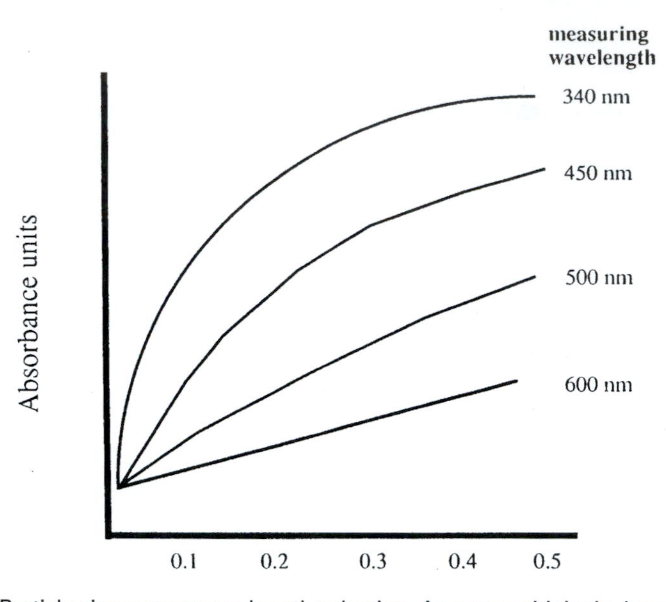

Figure 2. Particle size versus wavelength selection. Assay sensitivity is determined by the ability to detect early changes in aggregation (duplex formation ideally). A doubling in size of a particle of <0.1 μm when measured at 340 nm gives the greatest change in turbidimetric signal.

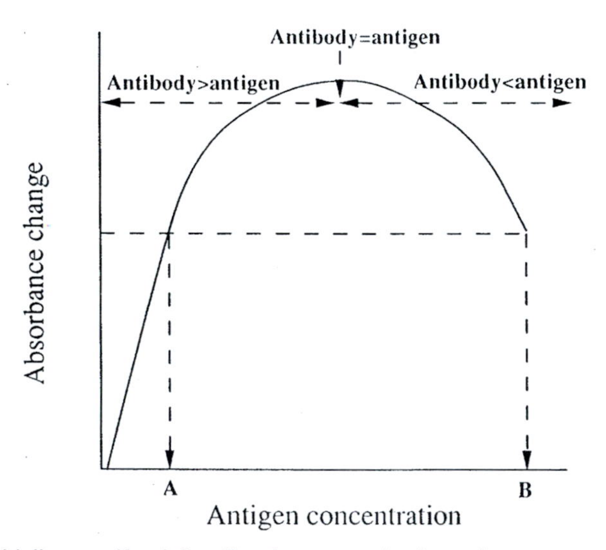

Figure 3. Heidelberger–Kendall calibration curve. As the antigen concentration increases, with a fixed amount of antibody, the scattering signal increases to a maximum and then decreases. This characteristic appearance is due to the changing antibody:antigen ratio. As the ratio falls below one, the excess in antigen causes a decrease in the aggregate size as two antigens begin to bind to each antibody molecule. This causes the problem of a single signal change being caused by two antigen concentrations A and B.

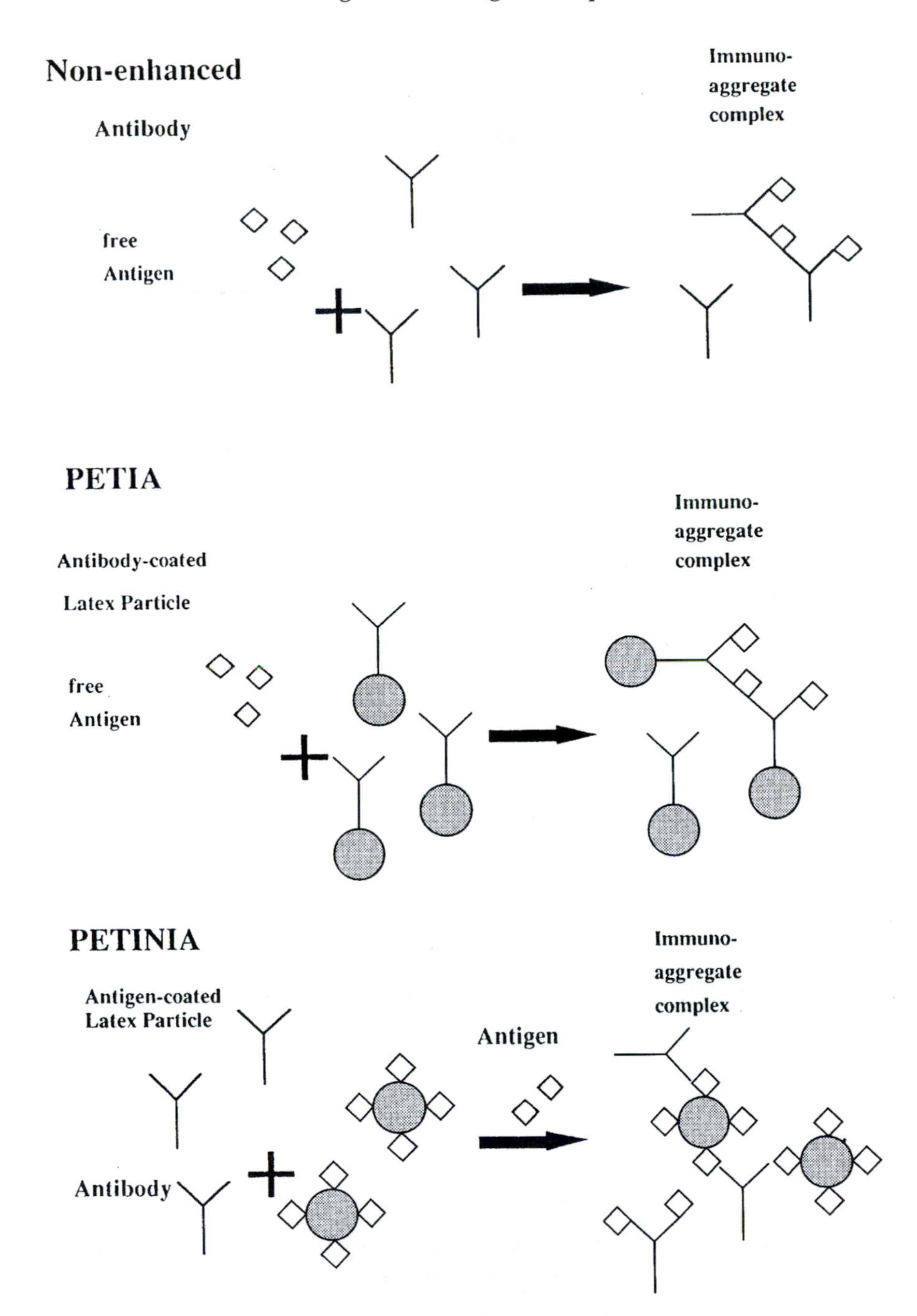

Figure 4. Assay formats (non-enhanced, PETIA and PETINIA). This cartoon represents the reactions that occur in the different assay formats.

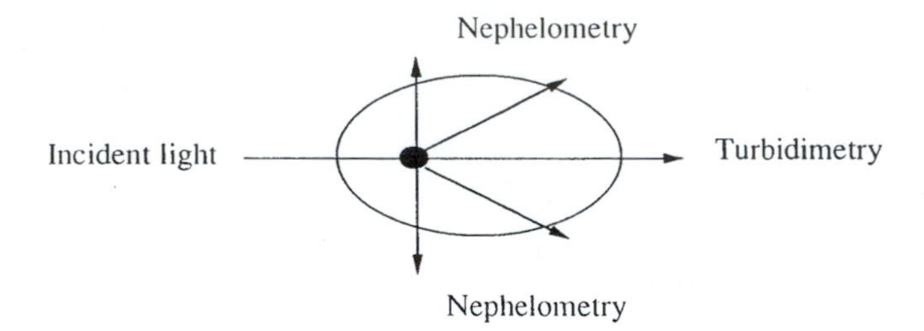

Figure 5. Turbidimetry and nephelometry: The arrows indicate the intensity of scattered light, in turbidimetric monitoring the loss of light passing through the reaction is monitored and in nephelometry the increase in light scattered away from the direction of the incident light beam.

decrease in the light passing through a solution and is a close equivalent of absorbance measurement, that is measurement of a decrease in signal against a high background. Nephelometry is the measurement of light scattered away from the angle of incidence of a light beam passing through a solution; this is analogous to fluorimetry, that is, the measurement of an increase in signal against a low background. The end result of this is that nephelometry should potentially provide a more sensitive detection system, although this has not always been achieved in practice due to greater interference from light-scattering species, such as lipoproteins, which are found in biological fluids.

Both turbidimetry and nephelometry are equally applicable to particle enhanced assay systems, but there is a further monitoring technique only applicable to particle based systems and that is particle counting (8). This is not readily available in most laboratories so will not be discussed further.

2.2 End-point versus rate measurement

The final choice to be made is whether to use end-point or rate measurement of the signal generated. In most cases, rate measurement is to be preferred as this enables an initial sample blank to be measured and multiple readings to be taken over the course of the reaction, thus improving precision of measurement (*Figure 6*). However, fixed interval (final minus initial) is undertaken rather than true rate. If a microtitre plate reader or a manual spectrophotometer is used, then end-point measurement may be the only option.

2.3 Analyte and assay format

For analytes with multiple epitopes found in the concentration range 50 mg/litre up to g/litre then direct immunoturbidimetry is the most appropriate analytical mode, for concentrations lower than this then particle enhancement is required. The different assay formats are described in *Figure 4*.

The most important considerations are what is the concentration range of the analyte and what is its molecular weight. For high sensitivities and low

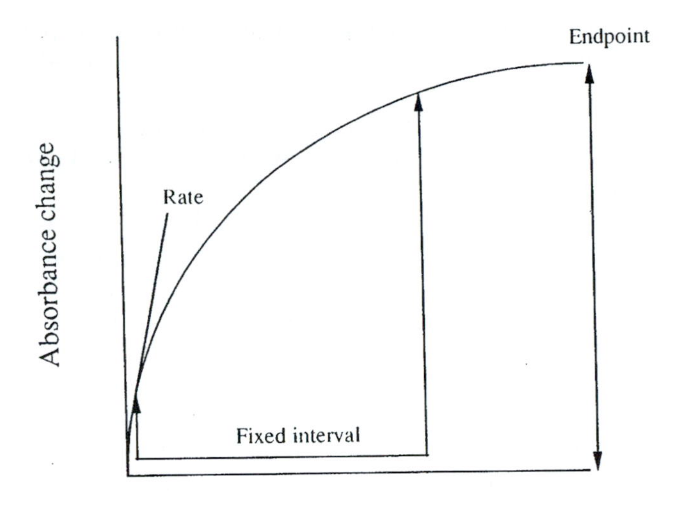

Figure 6. End-point versus rate measurement: End-point monitoring is the simplest form of signal measurement, simply waiting for the rate of reaction to reach zero after which accurate timing of the reading taken is not necessary. The problem with this is that there is no assessment of the contribution of sample absorbance to the overall signal at short time intervals enabling the rate of reaction to be determined, this provides a very precise monitoring system but requires sophisticated computing. Fixed interval measurement is a compromise that uses an early read after reagent and sample mixing, to determine the contribution of sample to the absorbance, and then a second later read from which the early read is subtracted.

molecular weight analytes then particle enhancement is the preferred option. The scattering signal can show, with careful selection of the appropriate particle, a 10- to 100-fold increase. Particles used include red blood cells, gold sols but most commonly latex particles (2–5,9). The classical assay formats are shown in *Figure 4* and, as can be appreciated, low molecular weight analytes will not be able to form an immunoaggregate unless conjugated to a particle to give a polyvalent haptenic complex, the free hapten can then be measured by its inhibitory effect on the agglutination caused by the antibody.

2.4 Antigen excess

One of main concerns with the use of all light scattering assays is to avoid antigen excess difficulties. During assay validation and optimization the limits of the assay performance can be selected with a knowledge of the likely pathophysiological range of the analyte of interest. In practice, the antibody concentration is optimized to give a wide enough working range for the assay, such that a signal change at the highest possible antigen concentration is still high. *Figure 3* shows that at point A and B there are equal signal changes, the concentration at A is high enough that there will be only a small percentage of sample higher than this point. The signal change at A is then used as a cut off, all signal samples with signal changes above this will be re-assayed following dilution. This signal change can be stored in the method protocol on most automated analysers as a Final Optical Density (FOD) check, and samples reaching this will be automatically flagged for dilution.

Once the assay is in use it is still necessary to be able to check that an exceptionally high sample has not been analysed. The reaction kinetics in the antigen excess region of the Heidelberger–Kendall curve are different to the antibody excess part, and some automated systems can monitor this. A more generally applicable approach is to take an early reading, as is the norm in fixed interval monitoring, in antigen excess samples (and in samples causing non-specific aggregation) the initial read can be dramatically elevated. Thus by careful analysis of the raw signal data antigen excess can be identified and the sample re-analysed on dilution.

2.5 Optimization strategies

An optimization strategy for a light scattering requires an understanding of the performance requirements for the assay in terms of which analyte is used; what is the desired sensitivity (detection limit), working range and equivalence point and finally what instrumentation is available. Often the latter choice is the most limiting, if a nephelometer or particle counter is not available, then turbidimetry with or without particle enhancement is the only choice. The general approach to developing both enhanced and non-enhanced assays is similar and the main points are summarized in *Table 1*.

2.6 Validation strategies

Validation of immunoaggregation assays follows essentially the same approach as is used for all immunoassays (*Table 2*), one just needs to be

Table 1. Optimization strategy for light scattering assays

What instrumentation is available?
What analyte is to be measured?
What detection limit and working range are required?
Select assay format: non-enhanced, PETIA versus PETINIA
Select a range of antibodies, either affinity purified or at least immunoglobulin fractions
Select a particle size and coupling chemistry
  Couple immunoglobulins to particles across a range of protein concentrations (0.5 to 4 mg/ml)
Select a suitable buffer
  PEG optimization
Optimize for signal change, detection limit and assay range by varying sample size and reagent concentrations

Table 2. Method validation for light scattering assays

Non specific binding (non-specific aggregation)
Detection limit
Working range and equivalence point
Imprecision
Calibration stability
Analytical recovery
Parallelism
Method comparison (with a non-light scattering methodology)
Interferences:

Spectroscopic:	icterus, haemolysis, lipaemia
Immunological:	rheumatoid factor, myeloma serum (high total protein), anticoagulants (particularly EDTA and citrate)

particularly aware of the problems of non-specific aggregation and rheumatoid factor interference causing an increase in light scattering. Human anti-mouse (sheep/goat/rabbit) antibodies can also cause similar problems. If interference from these sources is identified, then addition of an excess of free animal IgG can eliminate the problem. Non-specific aggregation from other sources can be a consequence of a variety of factors (serum proteins, lipaemic samples) and needs to be eliminated during assay optimization.

3. Non-enhanced methodology

Turbidimetric immunoassays have been used to quantify many serum proteins, some examples are given in *Table 3*. The methods have to be optimized in terms of reaction time, buffer pH, PEG concentration, pre-dilution or pre-treatment of samples and type of instrument used. In practical terms, a common phosphate-based buffer of neutral pH is quite suitable (an example is given in *Protocol 1*), the main variable being the amount of PEG added.

Protocol 1. Buffer recipe for phosphate-based turbidimetric assay

A. To obtain a 100 mM reaction concentration, a 340 mM sodium phosphate buffer is used:

The preservative used here is sodium azide at a concentration of 0.1%. For 1 litre volume weigh out:

- 39.107 g Na_2HPO_4
- 7.752 g NaH_2PO_4
- 1 g NaN_3

Make volume up to 1000 ml, pH = 7.5.

Protocol 1. *Continued*

B. Potassium phosphate buffer

- KH_2PO_4, 0.47 g
- $K_2HPO_4 \cdot 3H_2O$, 2.12 g
- NaC1, 7.43 g
- NaN_3, 1.0 g
- H_2O, make up to 1000 ml
- pH 7.3 ± 0.1

Add BSA and PEG to these buffers, at the concentrations required by your assay, prior to making up to the final volume as they are both hydrophilic molecules.
NB. Some detergents are quite acidic/alkaline and it may be better to add before the pH is measured and adjusted.

The signal changes and kinetics for such assays are indicated in *Figure 7*. It is unusual for a signal change to exceed 0.2–0.3 absorbance units or require more than 5 min for a reaction to near completion. An example of a detailed protocol for serum β2m assay on a Monarch centrifugal analyser is shown below, to give an indication of the reagent and sample proportions. Assay imprecision even with these relatively small absorbance changes is usually

Table 3.

Protein	Assay range	Detection limit (sample volume/ pre-dilution)	Buffer constituents (in the reaction mixture)	Antibody
Serum IgG	0–25 g/litre	0.17 g/litre (5 μl/1:20)	(a) + 60 g/litre PEG	Rabbit IgG fraction 1:50 dilution
Serum β2m*	0–20 mg/litre	0.5 mg/litre (50 μl/neat)	(a) + 80 g/litre PEG and 150 mM NaCl	Rabbit IgG fraction 1:41 dilution
Serum α1-antichymotrypsin	0–30 mg/litre	5 mg/litre (25 μl/1:3.5)	(b) + 70 g/litre PEG	Rabbit antiserum 1:41 dilution
Serum transferrin	0–10 g/litre	0.5 g/litre (10 μl/1:51)	(b) + 40 g/litre PEG	Rabbit antiserum 1:41 dilution
Urine albumin	0–100 mg/litre	5 mg/litre (20 μl/neat)	(a) + 40 g/litre PEG	Goat IgG fraction 1:50 dilution

*Required pre-treatment to precipitate serum proteins using 36% PEG.
Buffer (a) 100 mM sodium phosphate pH 7.5; buffer (b) 13 mM potassium phosphate pH 7.3 + 127 mM NaCl.

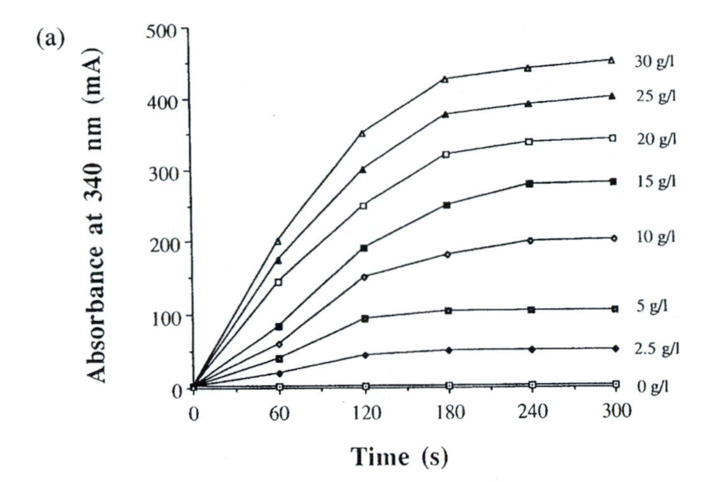

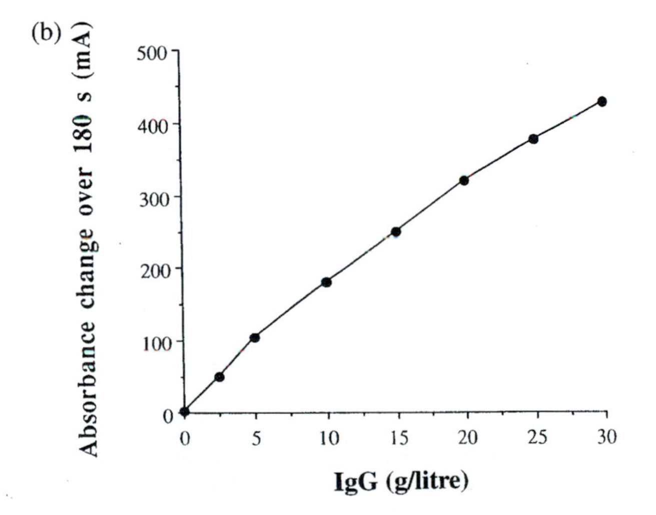

Figure 7. Non-enhanced calibration curves and kinetics. These curves are taken from an assay for serum IgG, the kinetics for the reaction with the different calibrators are shown in (a) and the resultant calibration curve plotted at 180 s shown in (b). The timings for (b) were selected on the basis that the reaction had reached near completion after 180 s and there was little to be gained by extending the read interval further.

good, i.e. <5% on automated spectrophotometers such as the Monarch, Cobas range etc. On these automated analysers fixed interval readings can be used with accurate timing of sample and antibody mixing, this is essential for precise results and this is why use of a manual spectrophotometer is inadvisable.

Protocol 2. Monarch protocol for a non-enhanced immunoturbidimetric assay for β2microglobulin

Sample volume (μl):	50	
Diluent volume (μl):	10	
Reagent 1 volume (μl):	150	(antibody diluted 1:41 in assay buffer A)
Diluent volume (μl):	10	
Reagent 2 volume (μl) :	Nil	
Total reaction volume (μl):	220	
Reaction temperature (°C):	30	
Monitoring wavelength (nm):	340	
Initial read time (s):	5	
Read interval (s)	60	
Number of readings:	9	
Total reaction time (min)	9	

3.1 Antibody selection

In general the antiserum, which can be goat, rabbit etc., is diluted directly into the assay buffer and added to the sample to start the reaction, which is usually monitored over a 5–10 min period at 30 or 37°C and at 340 nm wavelength. In selecting a suitable antibody at least three or four different sources (examples in *Table 4*) should be tried, in general a range of dilutions from between 1:10 and 1:100 should be looked at (*Figure 8*). In this example for the development of a serum IgG assay three different antisera have been compared at a common dilution (1:40). It can be seen that antibody B gives the greatest signal change, although not giving the highest equivalence point, whereas with antibody C too much signal would be lost. There are several ways of extending equivalence points as will be shown later and thus antibody B is selected for further investigation.

Having selected an antiserum, then the appropriate dilution needs to be identified by performing appropriate dilution curves as shown in *Figure 9*.

Table 4. Suppliers of specialist immunoreagents for immunoturbidimetry

Dakopatts	Glostrup, Denmark
Atlantic antibodies/Incstar Ltd	Minnesota, USA
The Binding Site Ltd	Birmingham, UK
Immuno Ltd	Sevenoaks, Kent, UK

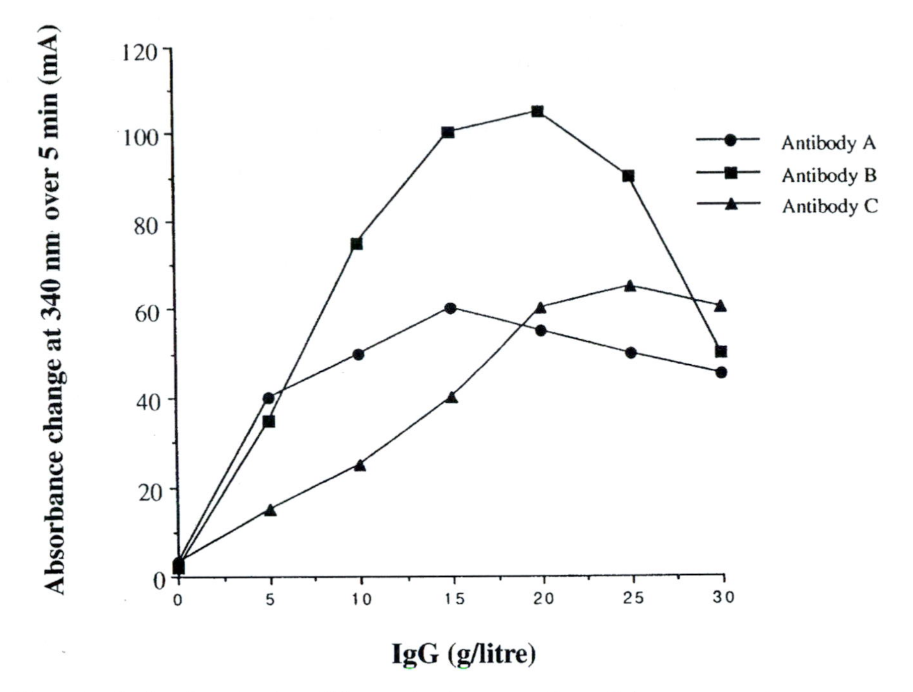

Figure 8. Antibody selection. When selecting an antisera it is necessary to compare at least three at one fixed dilution, usually in a generic buffer such as described in *Protocol 1*. Here three antisera for IgG are compared using serum-based IgG calibrators, and clearly antibody B gives the greatest signal change and widest assay range.

Using the same IgG assay example three different dilutions of antibody B are compared. The highest dilution does not give the required working range, the lowest dilution is too expensive, leading to the 1:40 dilution as the chosen compromise. Antibody dilution should be prepared fresh each day, as there can be variable deterioration in antiserum performance, diluted in the generally high PEG content buffers that are used (see *Figure 10*). Daily reagent preparation also necessitates daily preparation of calibration curves as this can differ between antiserum dilutions. This relative lack of reagent and thus calibration stability is one reason for conjugating antibodies to solid phases where there are significant improvements to be gained.

The use of intact immunoglobulin molecules does expose the assay to potential interference from rheumatoid factor, which can cause aggregation by cross-linking the FC regions of immunoglobulin molecules and thus generate a light scattering signal. Fab_2 fragments are one way of eliminating this difficulty, but control of the protease digestion step can be problematical with low yields of active antibody. The potential problem is, however, not as large as is commonly believed and can be eliminated by careful selection of buffer conditions. The range of immunoglobulin purification techniques that can be used are listed in *Table 5*, but these are really only required for the development of particle enhanced methods.

3.2 Choice of buffer and PEG concentration

Non-enhanced turbidimetric assays use a fairly narrow range of assay buffers, as can be seen from *Table 3*. Particle enhanced methodologies, which are discussed later, use a much wider range of buffer salts, pH values and other buffer constituents, such as detergents and added blocking proteins such as BSA. Antibodies should be selected in conjunction with the assay buffer. The buffer conditions can be of considerable importance in avoiding non-specific reactions, these can be both enhancement and inhibition of agglutination, and

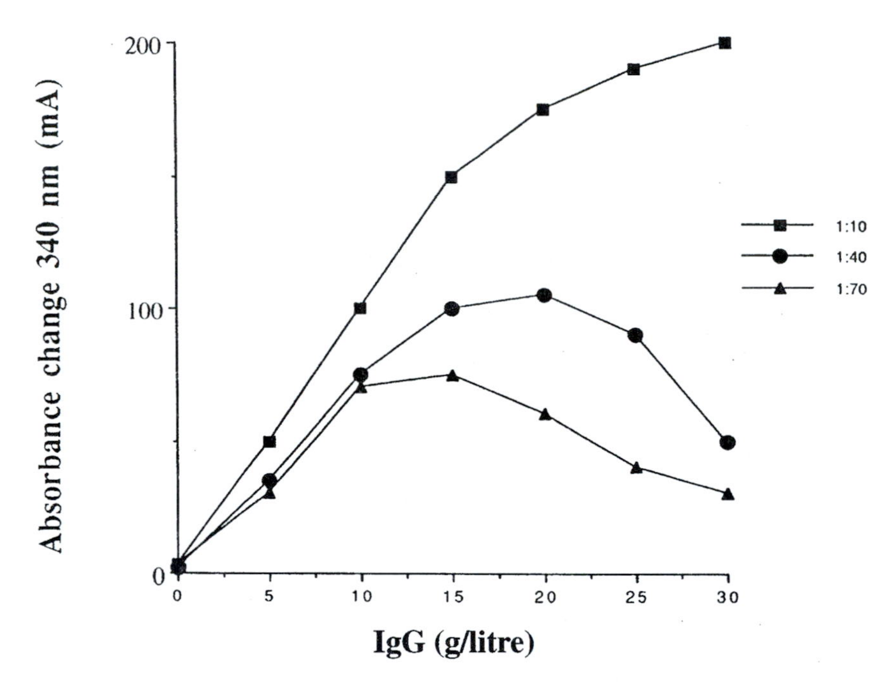

Figure 9. Antibody concentration and equivalence. Using the same IgG assay model a series of dilutions of antibody B are compared looking for maximal working range and signal change with an economic usage of the antiserum. Overall the 1:40 dilution offers the best compromise.

Table 5. Methods for the preparation of purified immunoglobulins and antibodies

Ammonium sulphate precipitation
DEAE chromatography on magnetic particles (e.g. Magnicel Scipac UK Ltd)
Protein A
Antigen-based affinity purification
Preparation of Fab_2 fragments by enzyme digestion

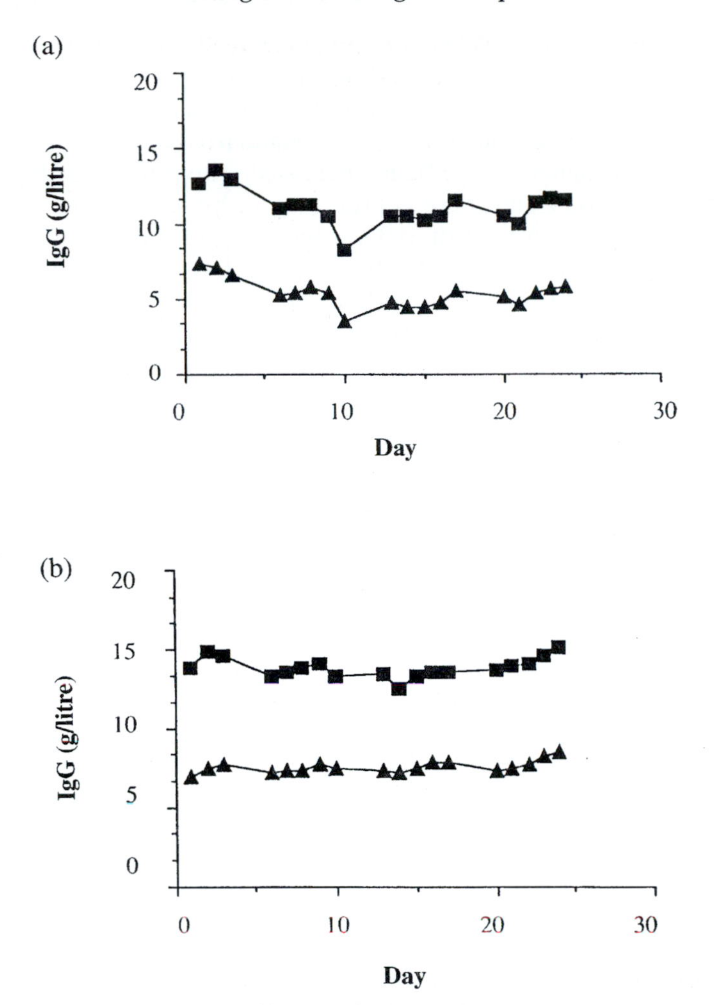

Figure 10. Antibody stability for IgG turbidimetric assay. Antibody was diluted in bulk and stored at 4°C between assays (a) or prepared fresh on the day of assay (b). Calibration curves were run over 20 days and the absorbance changes plotted. The signal change is more stable with reagents prepared on the day of the assay.

appropriate optimization experiments should be performed using the appropriate biological matrix with and without added analyte.

The use of PEG is very important, the commonest molecular weight PEG used is the 6–8000 size but other molecular weights can be used. PEG works by removing the water of hydration around the protein molecules, forcing them closer together and accelerating the aggregation reaction (10). With serum-based assays there is also the problem that PEG can enhance non-specific aggregation of serum proteins causing an increased blank rate, careful titration is thus required. When optimizing the PEG concentration it is important to investigate the kinetics of the reaction rather than just the overall signal change. The kinetics are significantly accelerated by the PEG, as shown in *Figure 11(a)* in which the 30 g/litre standard data alone is plotted, as well as the overall signal change being increased. The rate is starting to slow significantly after 180 s and there is little to gain by leaving the reaction to carry on, thus a reaction time of 180 s using 3% PEG in the final reaction mixture could be used. *Figure 11(b)* shows the additional advantage of PEG, producing an increase in the equivalence point from 15–20 g/litre up to >30g/litre. From these simple experiments, which can be performed in half a day, antiserum dilution and PEG concentration have both been selected.

3.3 Assay performance

Examples of assays that can be run on an automated spectrophotometer such as the Monarch analyser from Instrumentation Laboratories are shown in *Table 3*. The assays all use the same assay buffer but vary in the sample pre-treatment used: for IgG and α1-antichymotrypsin, it is a pre-dilution in saline; for β2-microglobulin, a PEG precipitation is necessary to gain sufficient sensitivity; for urine albumin, the sample is assayed neat. The remainder of the assay protocols on the instrument are very similar. Assay reaction times, signal change and imprecision within their respective working ranges are very similar, i.e. CV less than 5%.

This kind of approach can be used for assays within the concentration range of a few mg/litre up to g/litre but to increase sensitivity particle enhancement is required.

4. Enhanced methodology

Particle enhancement is essential for high sensitivity light scattering assays and for the measurement of haptens. There are a wide variety of particles available for this purpose, e.g. erythrocytes and metal sols, but the only really practical choice are latex particles (2–5,9). The first latex enhanced assay was published in 1956 by Singer and Plotz, a slide based test for rheumatoid factor (11). They and many others since have made serendipitous use of homogeneously synthesized particles, predominantly prepared from polystyrene. There are only very few particles that have been deliberately designed for use in light scattering assays, those from Litchfield *et al.* and Kapmeyer *et al.* being the most well known (12,13). Although these selectively designed particles do appear to offer advantages over commercially available latex, they are only available for use in the companies own assays (Dade International and Syva–Behring, respectively).

The advantages of particle enhancement in terms of signal change and assay sensitivity are clearly shown in *Figure 12* for two different assays, in

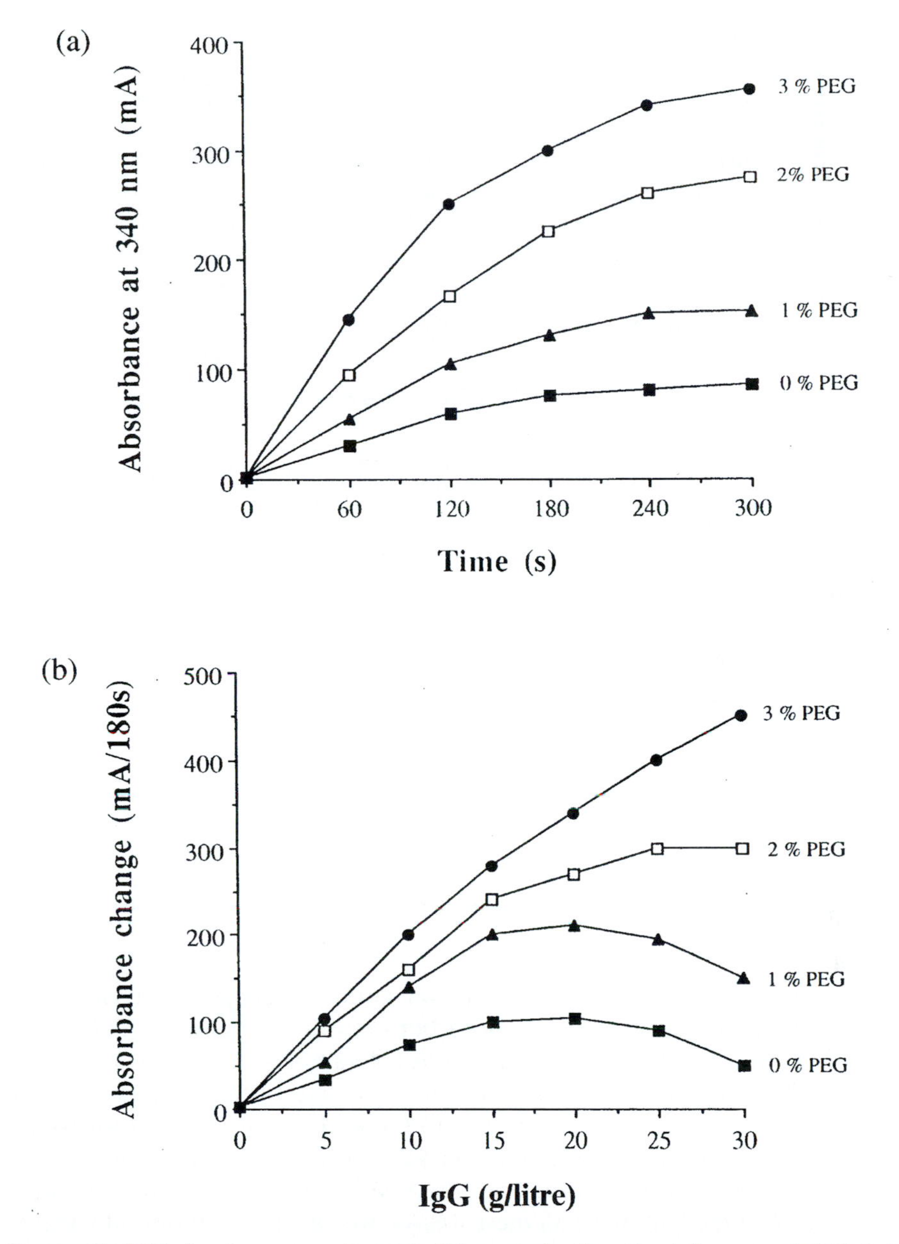

Figure 11. PEG titration non-enhanced. When evaluating the influence of PEG it is essential to study both reaction kinetics (a) and calibration curve (b). Here antibody B at 1:40 dilution is used to prepare calibration curves over the range of 0–3% (in the reaction cuvette) of PEG 6000. The kinetics show an enhanced rate of reaction with increasing PEG but the overall time to reach completion is not significantly changed. The calibration curve shows that PEG addition results not only in an increased signal change but also an extended working range.

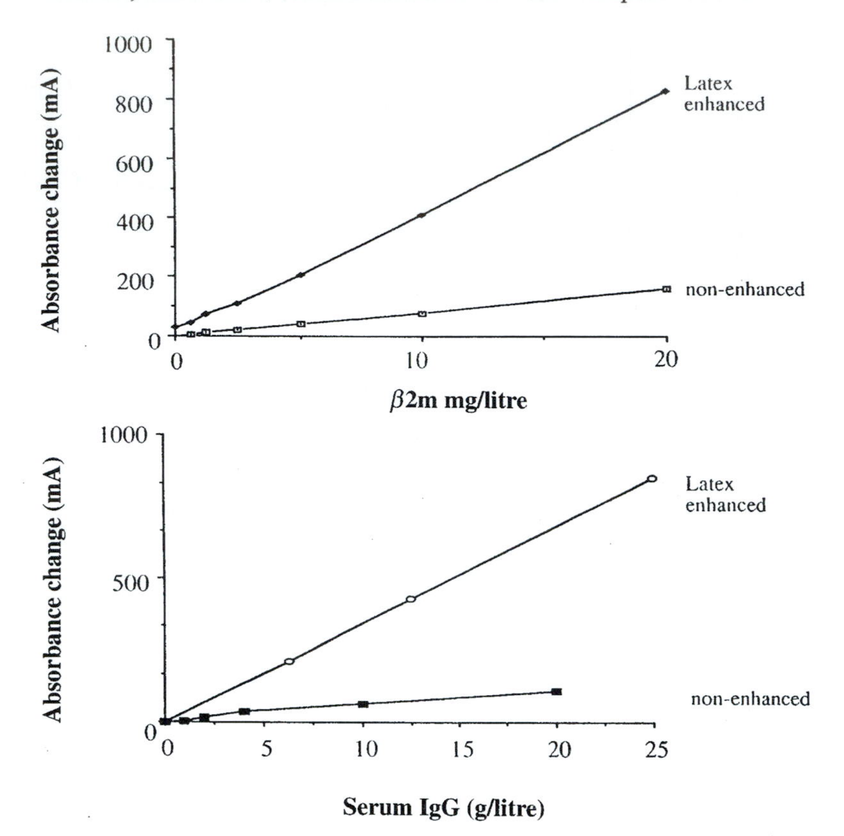

Figure 12. Enhanced versus non-enhanced calibration curves. Here are shown the differences in signal change for two assays that have been compared in both non-enhanced and latex particle enhanced formats with the same antisera used in both formats.

both cases the same antibody is used (conjugated or unconjugated). However, not only is the signal change enhanced but so is the antiserum stability and utility (i.e. more tests per mg of antibody). Examples of particle enhanced assays are shown in *Table 6*. All these assays have been developed in our laboratory, some proving more successful than others, the reasons for which we shall discuss later. In general it can be seen that there is a wider assay buffer pH range, less PEG used and the occasional use of blocking proteins such as BSA. One feature of all these assays, which use two different particle surfaces, is that none of them show interferences from rheumatoid factor, despite using intact immunoglobulins.

Protocol 3 shows an instrument protocol for a particle enhanced assay; there are many similarities with *Table 3*, but the main difference is that an enhanced assay requires two reagent additions. This is because a latex particle suspension is a colloid, which is only stable under well-controlled conditions

Table 6. Particle enhanced methods

Assay format	Matrix	Protein	Protein size	Assay range	Detection Limit	Ab loading (0.5% particle)	CMST particle size	Final buffer constituents in the cuvette
PETIA	Serum	Myoglobin	14000	0–0.5mg/litre	0.025 mg/litre	3.0 mg/ml (rabbit IgG)	40 nm	Buffer 2 + 0.3% BSA + 1.0%/PEG
PETIA	Serum	β2-Microglobulin	11800	0–20 mg/litre	0.42 mg/litre	1 mg/ml (rabbit IgG)	40 nm	Buffer 1 + 0.3% BSA, no. PEG
PETIA	Serum	Cystatin C	13000	0–10 mg/litre	0.027 mg/litre	0.25 mg/ml (rabbit IgG)	77 nm	Buffer 2 + 0.3% BSA + 2% PEG
PETIA	Urine	Retinol binding protein	21000	0–6 mg/litre	0.025 mg/litre	2 mg/ml (rabbit IgG)	40 nm	Buffer 2 + 0.3% BSA + 2% PEG
PETIA	Urine	α1-Microglobulin	33000	0–30 mg/litre	0.5 mg/litre	4 mg/ml (rabbit IgG)	40 nm	Buffer 1 + 1% PEG
PETIA	Urine	Albumin	66000	0–200 mg/litre	0.3 mg/litre	4 mg/ml (goat IgG)	40 nm	Buffer 1 + 1% PEG
PETINIA	Urine	Albumin	66000	0–200 mg/litre	0.1 mg/litre	2.0 mg/ml (albumin)	40 nm	Buffer 3 + 1% PEG
PETIA	Serum	Pre-albumin	55000	0–550 mg/litre	2.5 mg/litre	2.5 mg/ml (sheep IgG)	40 nm	Buffer 3 + 1.4% PEG + 0.01% Gafac RE610
PETIA	Serum	C-reactive protein	118000	0–120 mg/litre	0.60 mg/litre	2 mg/ml (goat IgG)	40 nm	Buffer 1 + 1% PEG

Buffer 1: 100 mM phosphate pH 7.5.
Buffer 2: 100 mM borate/KCl pH 10.0.
Buffer 3: 150 mM phosphate pH 7.8.

of pH, ionic strength, protein loading etc. otherwise it will self aggregate. Thus particle reagents are stored in low ionic strength buffers in which they can be stable for years. Low ionic strength buffers are, however, not suitable for serum-based assays where the high protein concentration would also cause non-specific aggregation, thus a compromise needs to be reached, and in general this is achieved by storing the particle reagent in a different buffer to the reaction buffer. Under the assay buffer conditions the colloid will be balanced on the edge of self agglutination which is why very careful reagent optimization is necessary and particle enhanced methodologies are less straightforward to develop than non-enhanced.

Protocol 3. Instrument protocol for a particle enhanced immunoturbidimetric assay for serum β2m

Sample volume (μl):	5	
Diluent volume (μl):	5	
Reagent 1 volume (μl):	160	(antibody particle reagent diluted 1:30) (assay buffer) (see Protocol 9)
Diluent volume (μl):	10	
Reagent 2 volume (μl) :	80	
Total reaction volume (μl):	260	
Reaction temperature (°C):	30	
Monitoring wavelength (nm):	340	
Initial read time (s):	5	
Read interval (s)	60	
Number of readings:	8	
Total reaction time (min):	8	

4.1 Choice of assay format

Both antigen and antibody can be conjugated to particles, for haptenic antigens covalent conjugation is essential as in order to achieve a viable assay strenuous washing is necessary to remove unbound hapten. However, for protein antigens, e.g. albumin, both approaches can be used and working assays developed, see *Figure 13*, after suitable optimization. Using an inhibition format can be extremely useful if a wide range of antigen concentration is to be expected, as is the case, for example, with urine albumin.

4.2 Selection of particle

There are a wide variety of latex particles available with different sizes, surface chemistries and colours, and there are several manufacturers to

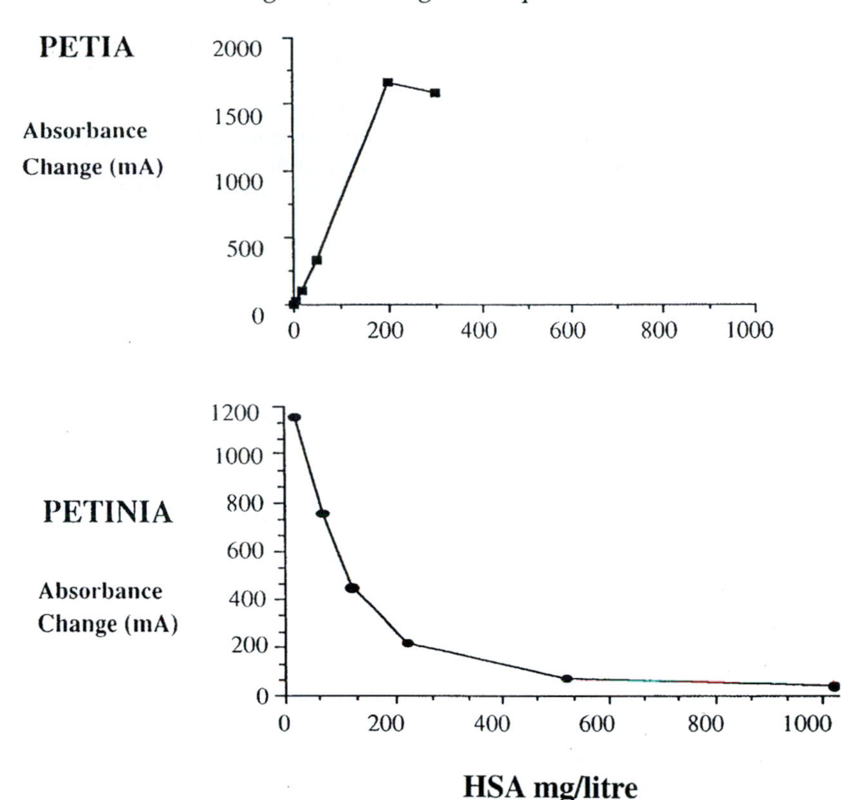

Figure 13. Direct versus inhibition of agglutination standard curves: Albumin example. Two assays for urine albumin were developed using the same antibody, in the PETIA the antibody is coupled to a 40-nm CMST particle and in the PETINIA purified human serum albumin is coupled to the same particle type. The signal change is comparable for both assay formats, but the difference in working range provided by the inhibition format is clear and its freedom from antigen excess difficulties apparent.

choose from (*Table 7*). The main choices to make are particle size and coupling chemistry.

4.2.1 Particle size

For most turbidimeters a small particle in the region of 20–150 nm is preferable, using a measuring wavelength of 340 nm. This size particle provides good scattering capability, remains in solution without mixing and provides a high surface area to facilitate reproducible coupling of protein. The main problem with such particles is that they are difficult to wash at the end of a coupling procedure and very high 'g' forces are required. The effect of different particle size on assay signal change and sensitivity can be seen in *Figure 14*, using a PETIA for albumin as an example. The 40-nm particles

Table 7. Suppliers of latex particles

Bangs Laboratories	Carmel, Indiana, USA
Seradyn Inc.	Indianapolis, Indiana, USA
Polymer Laboratories	Church-Streton, UK
Duke Scientific Corporation.	Palo Alto, California, USA
Rhône Poulenc	
Dynal	Skoyen, Oslo, Norway
IDC	Portland, Oregon, USA

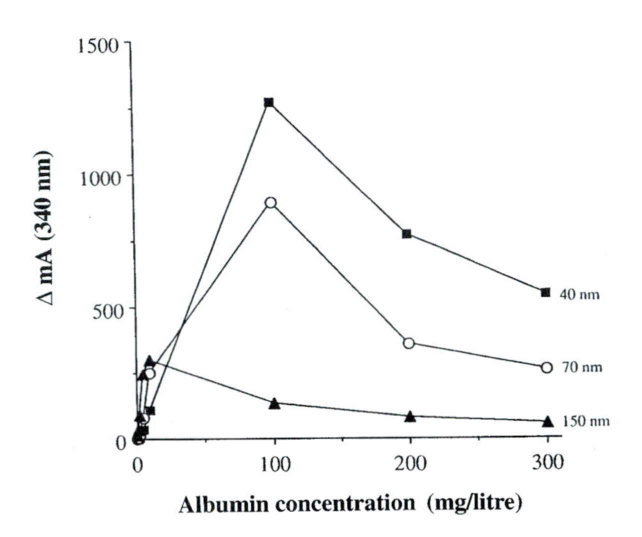

Figure 14. Optimization of particle size: Three different sized CMST particles were coupled at equivalent protein loading to a common anti:human serum albumin IgG fraction, the 40-nm particle clearly shows the greatest signal change.

generate a much more sensitive reagent with an extended working range in comparison with the larger particles.

4.2.2 Particle surface

There is a wide range of particle surfaces available, which can differ in the base polymer (polystyrene versus polyvinylnapthalene) to the co-polymers that are introduced to produce either stable active surfaces such as vinylbenzyl chloride (chloromethylstyrene) or carboxylated surfaces. The selection of which surface to choose can be very important, as even with the same coupling chemistry there can be significant differences in reagent performance, as shown in *Figure 15.* This shows two reagents prepared using chloromethylstyrene coupling chemistry but having different bulk polymers, there is a clear difference in the effects of different antibody loadings, pH effects and serum non-specific aggregation. Both particles could be used to develop assays, the one we selected in the end was the polystyrene surface because it was readily available commercially.

4.3 Coupling chemistry

Selection of a coupling chemistry can be complicated. There is little experimental evidence but the general assumption is that covalent coupling provides a more robust reagent; however, adsorption chemistry can provide perfectly usable reagents, it is just likely that the long-term stability will be less. Preparation of adsorbed reagents is nominally simpler (see *Protocol 4*), and is perhaps worth exploring first if you have little experience of covalent coupling chemistries. For any coupling technique to be reproducible it is essential that the antibody contains no aggregates, the buffers are sterile and freshly made, in high quality de-ionized water, and the particle and antibody solutions are mixed together slowly. Training courses in the preparation of latex particle based reagents are available through Bangs Laboratories, and Seradyn Inc. have produced a very useful guide to reagent preparation, based on their own experience, that is available for purchase (3).

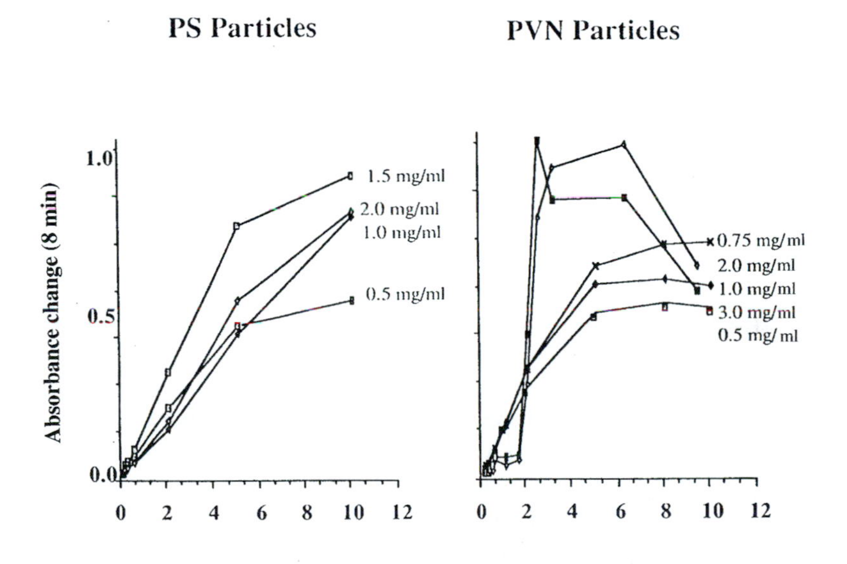

Figure 15. Selection of particle surface; Two different preactivated (CMST) particles, one predominantly prepared from polystyrene and the other a co-polymer with polyvinylnapthalene, were conjugated to differing amounts of an IgG fraction to human cystatin C. Both particles demonstrate suitable calibration curves but with differing optima for protein loading and (not shown here) different sensitivities to pH and non-specific aggregation. Both particles can be used to develop assays but different optimal conditions will be required (assay buffer for PS was buffer A + 1% BSA; PVN used buffer A + 1% BSA and 3% PEG).

Protocol 4. Protocol for adsorption of proteins to latex particles

Method

Dilute 1% latex particles in 25–50 mM MES buffer pH 6.1, and in a separate tube the protein solution in the same buffer (0.5–2.0 mg/ml). Optimization of protein loading with adsorption is generally less precise, most workers try to saturate the surface of the particle and thus leave a small amount of antibody free in solution. A range of protein loadings should be explored. Mix by adding equal volumes of protein solution to the latex solution, at room temperature, addition being slow and dropwise.

The reaction will essentially be complete in a few minutes but it is preferable to leave mixing over night and then the reagent is ready to use. Opinions vary as to whether a bulk protein such as BSA should be added to block remaining sites on the latex. Empirical optimization will be required but approximately 4 mg/ml BSA has been used in some instances.

4.3.1 Adsorption

We have preferred to develop covalent coupling protocols, but have performed some evaluation of adsorption protocols in comparison with covalent conjugation and the use of Protein A surfaces as potentially a universal surface. *Figure 16* shows a comparison of the standard curves generated by reagents prepared using the same amount of anti-β2-microglobulin immunoglobulin fraction immobilized on the surface. Adsorption was carried out according to *Protocol 4* with and without prior ion-exchange pre-cleaning of the particle surface; direct IgG and Protein A conjugation used CMST coupling. Protein A has been suggested as a potential universal capture agent for particle based assays (14); as shown in this experiment, it can produce a particle reagent with, at least on initial evaluation, suitable sensitivity. However, we have no experiences of developing and validating assays using this approach. Ion-exchange pre-cleaning is considered necessary by many authors as the particle surface often contains detergent, monomers etc., which can at least potentially interfere with the adsorption of protein to the surface. There is a difference in behaviour between the two treatments at the two pH values investigated (data not shown), but both give respectable standard curves at pH 10.0 that are very similar to the covalent coupled reagent and all are better than the Protein A based reagent.

4.3.2 Covalent coupling

This is most commonly achieved using carbodiimide chemistries in combination with a latex particle with exposed carboxyl groups on its surface,

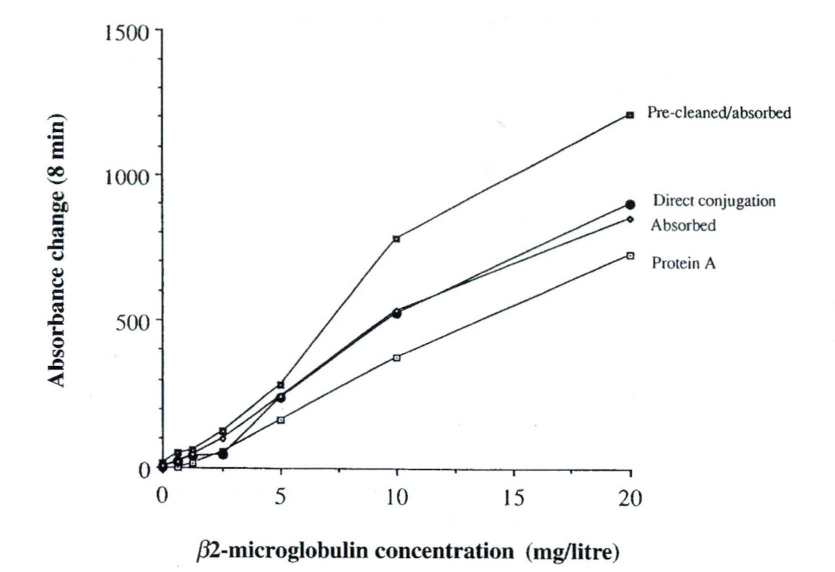

Figure 16. Covalent versus adsorptive versus Protein A conjugation: An IgG fraction to human β2m was coupled (at 1 mg/ml protein loading) to 40-nm CMST particles in four different ways; (1) direct conjugation; (2) preparing a Protein A conjugated particle (1 mg/ml loading) then adding the IgG fraction to this capture agent; (3,4) adsorbed to particles following blockage of their CMST groups with excess glycine with or without pre-cleaning of the latex with a mixed bed ion-exchange resin. The Protein A 'universal particle' gave the lowest signal change, but there was very little difference between the preparations and all could have been used to prepare assays. The reaction conditions used varied widely and in particular the ion-exchange pre-cleaned particles showed an enormous sensitivity to pH-induced non-specific aggregation.

which are first of all activated by reaction with a carbodiimide reagent and then reacted with protein. Our main choice has, however, been to use particles with stable preactivated surfaces of chloromethylstyrene, which can be directly conjugated with protein. We have explored other chemistries, such as aldehyde conjugation (also a stable preactivated surface), derivatizing carboxyl groups to generate amide groups for use with glutaraldehyde and the use of heterobifunctional linkers, but we have returned to the use of CMST latexes in the end.

Carbodiimide coupling

A protocol for covalent coupling using this approach is shown below *Protocol 5*, this is relatively straightforward but does require optimization of the carbodiimide concentration with different carboxyl content latexes. Carbodiimide chemistries are relatively aggressive and we have found that although they can give rise to satisfactory reagents, they are harder to optimize and give rise to large increases in particle size following protein conjugation, sometimes two- to three-fold. Such an increase in particle size suggest significant cross-linking of particles and this alteration in size will reduce the ultimate sensitivity of the assay.

Protocol 5. Carbodiimide coupling to carboxylated latex

Preactivation (two step) procedure: Carboxylated latex only.

I. Preactivation step

1. Pipette into microcentrifuge tubes in the order given:
   - 100 μl of 500 mM MES pH 6.1 buffer (50 mM final)
   - 100 μl of 10.0% solids stock microparticles (1.0% solids final)
   - 230 μl NHS solution (100 mM final)
   - 230 μl EDAC solution (1–2 mM final)
   - 240 μl deionized water to make 1.0 ml final volume
2. Mix tubes at room temperature on a mixing wheel or other device for 30 min.
3. Centrifuge and discard supernatant. Resuspend particles with 1 ml 50 mM MES buffer, pH 6.1. Centrifuge again, discard supernatant.

II. Coupling step

1. Resuspend pellet by adding the following and sonicate:
   - 50–100 μl 500 mM MES buffer (25–50 mM final)
   - 250 μl protein (antibody) solution (1–8 mg/ml final)
   - Water to make 1.0 ml final volume
2. Mix tubes at room temperature on a mixing wheel or other device for 1 h.

Chloromethylstyrene

These particles are less widely available but Bangs Laboratories does have a reasonable range of sizes. When using preactivated surfaces it is, however, necessary to remember that they have a limited shelf life compared with carboxylated particles, probably less than 6 months. This is more gentle coupling chemistry, a protocol for which is given below. This chemistry gives usually only a 20–50% increase in size following conjugation (*Protocol 6*).

Protocol 6. Coupling to CMST particles

Method

1. Dialyse the immunoglobulin fraction into a large volume of dialysis buffer (3–5 litre) for 24 h at room temperature.
2. Spin down the dialysed antibody and measure the optical density of the supernatant at 280 nm to calculate the concentration of the total protein content.
3. Dilute the antibody in coupling buffer to the appropriate concentration.
4. Prepare a 1% particle solution in the coupling buffer.
5. Add equal volumes of the diluted particle to the antibody solution by slow and continuous mixing at room temperature.
6. The coupling reaction is allowed to continue overnight at 37 °C in a shaker/incubator to ensure continuous mixing.
7. After overnight reaction, the particle reagent is centrifuged in a high speed centrifuge, the supernatant removed and the pellet washed by one of the wash procedures.

4.3.3 Washing particle reagents

Most coupling reactions are not 100% efficient and it is thus necessary to remove the unbound antibody/antigen. This is most commonly achieved by centrifugal washing as described in *Protocol 7*. The problem with this is that this results in precipitation of the colloid as a pellet, which can be difficult to disaggregate for ultimate use in an assay, hence the need for ultra-sonic disaggregation, which is difficult to control because different antibodies show differing robustness towards ultra-sonic energy. One way of avoiding this is to use tangential flow diafiltration using systems such as that supplied by Microgon. This enables washing without sedimentation and can result in higher particle reagent yields, see *Protocol 7*. The only problems with this approach are that it requires an additional piece of equipment (albeit much cheaper than an ultra-centrifuge) and that only one reagent can be washed at any one time.

Protocol 7. Washing latex particle reagents

I. Centrifugal washing

1. Centrifuge particles at 40000 g for 45 min, remove and store supernatant for protein assay.

Protocol 7. *Continued*

2. Add twice the reaction volume of the wash buffer, 50 mmol/litre glycine pH 7.4 containing 0.05% GAFAC, and resuspend the particle pellet using a Pasteur pipette.
3. Centrifuge particles at 40 000 g for 45 min, remove and discard supernatant.
4. Repeat steps 2 and 3 at least three more times.
5. Finally resuspend particle pellet in half the initial volume of storage buffer, e.g. 500 mmol/litre glycine pH 7.4, 0.1% sodium azide.
6. Sonicate particle suspension on ice, e.g. at 20 Hz on an MSE soniprep, for 2 × 1 min with a pause in between.

II. Tangential flow diafiltration protocol

During constant volume diafiltration the latex particle concentration remains constant whilst free protein passes through the membrane and is removed. The minimal latex volume that can be washed in the Microgon (Laguna Hills, CA, USA) system is about 10 ml and the smallest latex that can be washed at present is 100 nm.

1. Assemble the system as shown in *Figure 17* using a Micro/MiniKos Module with a pore size of 300 kD. The diafiltration membrane is first washed using 10–20 column volumes of buffer.
2. With the system filled with wash buffer add the latex particles to the reservoir as wash buffer is removed, to maintain a constant volume, and wash using 10–20 column volumes.
3. Recover the washed latex by draining the system, additional particle recovery can be achieved by backwashing.

4.4 Evaluation of particle reagents

All conjugations should be evaluated for coupling efficiency and the easiest way to do this is to measure the supernatant protein concentration in the first wash step, efficiencies should be greater than 80% (see *Protocol 8*). The other routine assessment is particle size and it does not have to be carried out using a sophisticated particle size analyser. The easiest technique is described in *Protocol 8*, measuring the ratio of size at two different wavelengths. The final assessment is clearly immunoreactivity and titre of the reagent.

Protocol 8. Assessment of particle reagents

1. Assessment of particle monodispersity

The particle reagent is diluted in 5 mmol/litre glycine, pH 7.4, and its absorbance measured at 340 and 600 nm. With absorbance at 340 of approximately 1, the 340/600 ratio is determined. The actual ratio varies depending upon the starting particle size but for 40-nm particles it should be >10.

2. Coupling efficiency

Measure the protein concentration in the first supernatant and in the starting antibody solution using a protein method not interfered with by detergents, e.g. Pierce BCL, and determine coupling efficiency by calculating the ratio of initial minus supernatant divided by the initial concentration.

3. Agglutination signal

Using a starting assay protocol, measure the signal change at zero antigen and at the top of the calibration curve, this should be consistent from lot to lot.

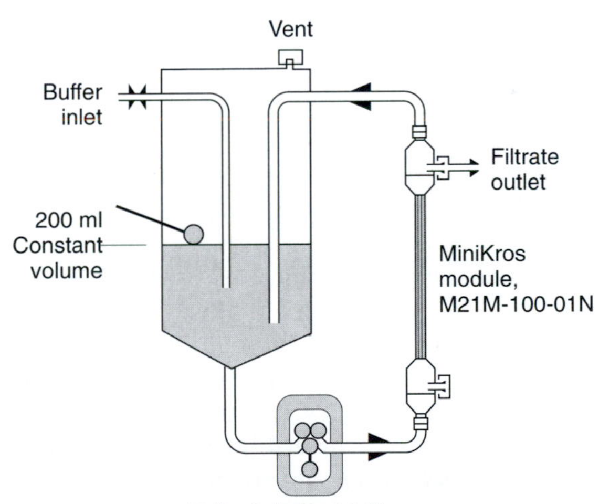

Figure 17. Tangential flow washing of latex particles. This technique of constant volume diafiltration is best used with relatively large-scale particle syntheses (>10 ml), but it provides a gentler washing system than centrifugal washing and can be used for particle synthesis as well.

Table 8. Equipment required for particle enhanced immunoassay

Ultra pure water
Incubator shaker
High speed centrifuge or dialysis/tangential flow equipment, e.g. Microgon
Ultra-sonic probe
Spectrophotometer

Whether adsorption or covalent coupling is used there are several important pieces of equipment that are necessary (see *Table 8*); high quality de-ionized water is vital to prevent non-specific agglutination of latexes during protein coating; a high speed centrifuge and an ultrasonic probe are essential for washing coupled latex. It can also be very important to use the sodium rather than the potassium salt of the buffer used, as this can cause less non-specific aggregation during conjugation. One important advantage of covalent coupling is that higher detergent concentrations can be used, this is very useful in preventing non-specific agglutination, as the detergent, particularly an anionic one such as SDS or GAFAC RE610, adsorbs to the surface of the particle giving rise to a high negative charge.

4.5 Antibody selection and optimization

Careful optimization of the amount of protein conjugated to the particle is vital, too much protein will destabilize the colloidal suspension and more reagent instability will result. Too much protein can also cause reduced immunoreactivity of the conjugated antibody due to steric hindrance. Optimization is an empirical process and examples of the kind of concentrations used and their influence on the resultant signal change and standard curve are shown in *Figure 15*. For the PS particles, the optimal loading was 1.5 mg/ml and for the PVN particles, 0.75 mg/ml. Both particles show that the maximum loading is not necessarily the best for steric hindrance and in particle instability occurring.

The loading required can vary significantly as shown in *Table 6*, and there can be a further choice to make as the higher the protein loading, the less stable the resulting colloid and the more difficult it will be to make a reproducible reagent. For the assays in *Table 6* with protein loadings of greater than 3 mg/ml we have had great difficulties in reproducibly manufacturing the reagents, so this is probably the upper limit for this kind of particle (size 40–70 nm CMST). The purity of the protein then becomes very important and if there is no other source of antibody it may be necessary to undertake affinity purification of the immunoglobulin fraction on an antigen column. This increases the complexity (and cost) of the reagent production but will result in the possibility of using a lower protein loading and a more reproducible synthesis and stable reagent.

Having optimized the amount of protein coupled to the particle surface it is then necessary to optimize the amount of latex that is added to the reaction. The high scattering induced by latex particles results in a high initial absorbance (the absorbance of the diluted particle reagent alone without added sample). Depending upon the path length of the cuvette used and the optical limitations of the spectrophotometer, it is best to start with an absorbance of approximately 0.5. This will allow a large signal to be generated (often 1.0 absorbance units) without reaching significantly into the non-linear part of the spectrophotometer's optical range. If a higher working range than usual is required, e.g. for pre-albumin (see *Table 6*), then more latex (and thus antibody) can be added by using a higher wavelength such as 405 or 450 nm, more latex is required to reach a starting absorbance of 0.5. Although the use of higher wavelengths will enable a wider working range, there will be a commensurate loss in assay sensitivity as predicted by *Figure 2*.

4.6 Buffer optimization

As mentioned previously, particle reagents require storage under low ionic strength conditions, but assay buffers need to have high ionic strengths to minimize non-specific aggregation effects. Some of our generic buffers for these purposes are described in *Protocol 9*. In general, we have found that all assays can be found to work in either of the two reaction buffers but there has needed to be further addition of PEG or BSA in some cases.

Protocol 9. Buffer recipes of preparation and use for particle enhanced immunoassays

A. Dialysis buffer for CMST latex particles

Needs 3 litres of 15 mM SODIUM phosphate buffer pH 7.4. (There seems to be a problem if potassium phosphate is used, as the particle reagent aggregates and gives a much lower and unacceptable ratio.)

- Dissolve

  5.176 g Na_2HPO_4

  1.026 g NaH_2PO_4

  in 3 litres distilled H_2O and check pH = 7.4.

B. Coupling buffer for latex particles

15 mM Na phosphate buffer pH 7.5 containing 0.05% GAFAC. The sodium phosphate buffer is the same as that used for the dialysis buffer with the addition of GAFAC. NB. Ensure to add the GAFAC ***BEFORE*** testing pH, as the detergent is acidic and will change the pH even at this low concentration.

Protocol 9. *Continued*

1. GAFAC (Gafco, Wythenshawe) prepare a 10% stock solution for ease of pipetting.
2. For 1 litre weigh out:
   - 1.725 g Na_2HOP_4
   - 0.342 g NaH_2OP_4
   - 5 ml GAFAC (10% solution)
3. Dissolve in 500 ml of distilled water, pH to 7.5, then make the volume up to 1 litre.

NB. When preparing the 10% stock solution of GAFAC pre-warm the neat GAFAC as this makes it less viscous and easier to handle.

C. Storage buffer (500 mm glycine pH 7.5)

1. Dissolve
   - 18.767 g glycine
   - 5.0 ml sodium azide (stock 10% solution) + 0.5% GAFAC

   in 250 ml of distilled water and adjust pH to 7.5 (add 2.5 ml of 10% GAFAC before adjusting the pH).
2. Make volume up to 500 ml.

D. Wash buffer (50 mm glycine pH 7.5)

For use when washing the Ab:Pr post coupling.

- Dilute the storage buffer 1:10 to give 50 mM glycine pH 7.5.

E. Diluting buffer (5 mm glycine pH 7.5)

For diluting the Ab:Pr for use in the assay.

- Dilute the storage buffer 1:100 to give 5 mM glycine pH 7.5.

F. Assay buffer: 340 mm borate/KCl buffer pH 10.0 (also known as Clark and Lubs solution)

Recipe below is for the 'base' borate/KC1 buffer. Additions of PEG and/or BSA, detergents can be made after to suit the assay. (NB. Certain detergents are acidic/alkaline and may need to be added before the pH is checked as they may alter it even when present at low concentrations.)

1. For 1 litre volume weigh out:
   - 21.022 g boric acid
   - 25.347 g KC1
   - 1.0 g sodium azide

2. Dissolve in 500 ml of distilled water (you may need to warm it up a little), adjust pH to 10 with 1.0 M NaOH (you will need quite a lot, e.g. 5 ml or so).
3. Add BSA/PEG/detergent to this stock solution at the concentrations required by your particular assay.
4. Make volume up to 1000 ml.

4.6.1 Non-specific aggregation

Non-specific aggregation is a major problem with particle enhanced assays. As discussed previously the colloidal suspension of protein-covered latex is placed in an assay buffer of high ionic strength for the immunological reaction to occur. Under conditions of low ionic strength, the charge repulsion of the latex particle surfaces not only reduces non-specific aggregation but also is sufficiently large so as to inhibit antibody induced aggregation. A reaction buffer has to be found that not only inhibits non-specific aggregation but also enables the antibody induced reaction to occur. The effect of different buffers and buffer pH values on one antibody particle system is shown in *Figure 18*. Only the borate buffer (buffer F from *Protocol 9*) at pH 10 reduced the serum induced non-specific aggregation to negligible levels (all the measurements

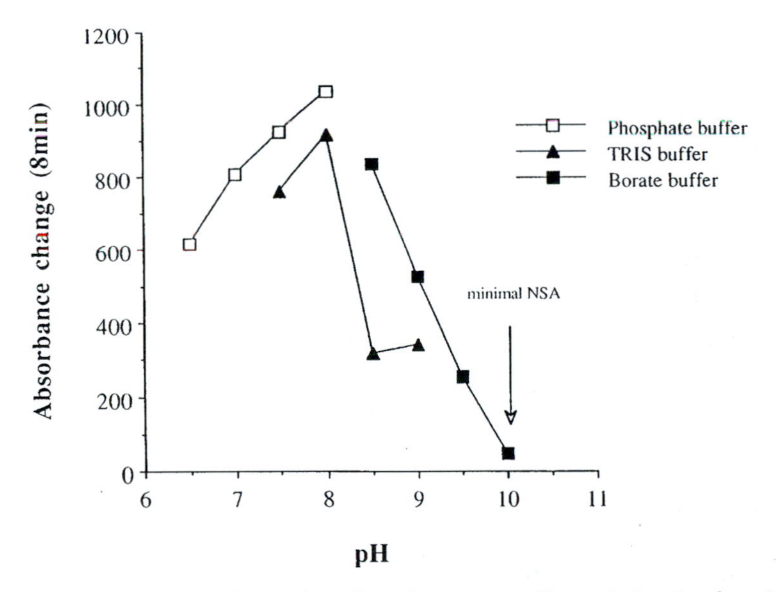

Figure 18. Effect of pH: This figure describes the non-specific agglutination found with human serum added to a latex preparation of anti-human β2m. Using a range of different buffers and pH values the minimal non-specific aggregation was found using pH 10.0 borate buffer. In a parallel series of experiments, the specific agglutination signal was also monitored and gave a suitable magnitude of response in this buffer. Different particle:antibody conjugates will show different non-specific aggregation profiles.

were carried out in the absence of antigen). Further experiments were then required in parallel to demonstrate that this buffer and pH did not inhibit the immunological reaction as well. Other particle:antibody conjugates may show very different pH optima and full optimization is necessary for each one.

4.6.2 PEG

As a consequence of the large signal generated in particle enhanced assays there is less need to add PEG to enhance the signal change. In general, an antibody loading on the particle is selected that gives good long term stability (i.e. <3 mg/ml) and this will give the maximum aggregation rate compatible with a stable reagent. PEG can be added to the assay buffer and its effect investigated on the agglutination signal. At the upper range of these protein loadings PEG can cause non-specific agglutination at concentrations above 1% in the reaction mixture, but can still act to increase the rate and extent of aggregation as shown previously for non-enhanced assays. In some cases it is possible that PEG can be used to help reduce the protein loading on the particle whilst maintaining the signal change.

In general it is always worth exploring PEG addition as described previously for non-enhanced assays, although the concentrations used should be a lot lower (nearer 1% in the final reaction mixture). The kinetics of particle based assays are fast, as shown in *Figure 19*, although reactions reach end-point in roughly the same time as non-enhanced assays. If a suitable signal change cannot be generated, to give an assay with the required sensitivity, then further purification of the antibody, e.g. affinity purification on an antigen column, will be necessary so that the amount of active antibody on the particle surface can be increased.

4.7 Assay performance

Examples of assays that can be run on an automated spectrophotometer such as the Monarch analyser from Instrumentation Laboratories are shown in *Table 6*. Assay reaction times, signal change and imprecision within their respective working ranges are very similar, i.e. CV less than 5%, as found for the non-enhanced assays at the higher antigen concentrations. One advantage of covalent coupling of antibodies to solid phases is the great enhancement in their stability. As shown in *Figure 20*, it is possible to achieve reagent and calibration stabilities of over a year, with significant savings on calibrators and reagents.

Due to the larger signals produced in combination with the rapid kinetics it is possible to adapt particle enhanced assays for use in a microtitre plate format (10). Using an end-point assay format, good signal changes and imprecision can be achieved, as shown in *Figure 21*, using the method described in *Protocol 10*. This opens up the possibility of using this kind of technology in the absence of a fully automated spectrometer.

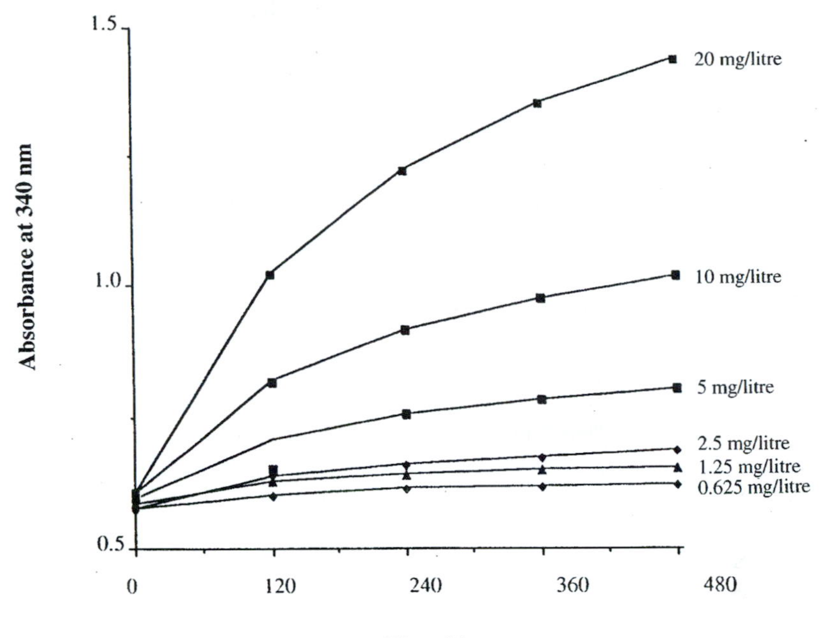

Figure 19. Reaction kinetics: These are the reaction kinetics for a serum β2m assay using 40-nm CMST particles. The kinetics are very similar to non-enhanced assays, simply giving a larger signal change.

Protocol 10. Adaptation of the serum β2M assay for use on a microtitre plate system

The Microtitre plate reader used for these experiments was a Molecular Devices V Max microtitre plate reader, linked to an Archimedes BBC computer; wavelength filter available was 405 nm. The maximum capacity for the plate wells was 250 μl. Volumes were all adjusted accordingly to attempt to keep all concentrations in the reaction mixture as near to the original formula as possible

- Sample volume = 3 μl
- Assay buffer volume = 73 μl
- Ab:Pr reagent volume = 160 μl
- Total volume = 250 μl
- Reaction time = 15 min

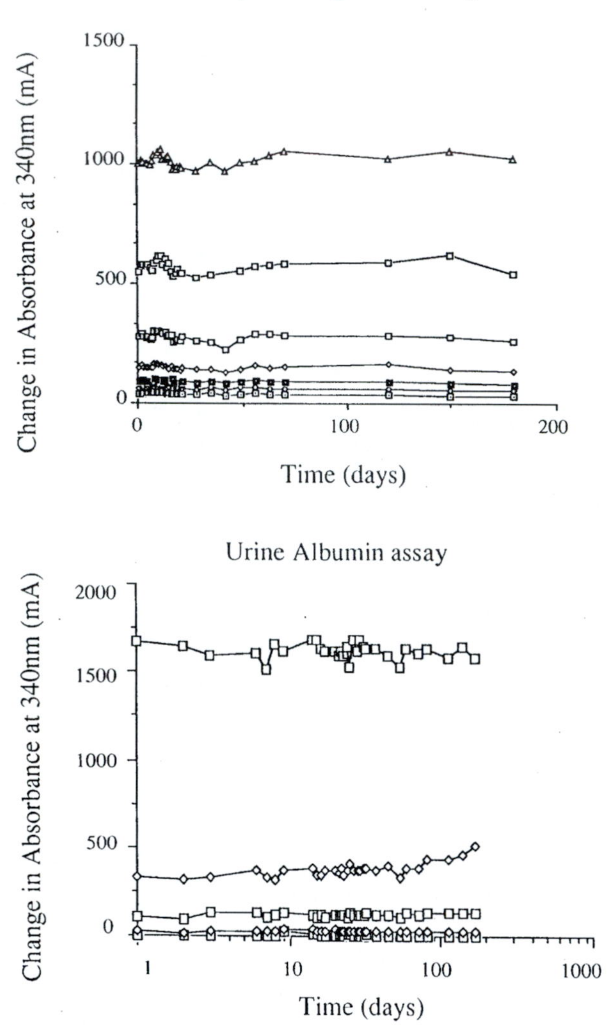

Figure 20. Reagent stability: Conjugating antisera to solid surfaces greatly enhances their stability. These examples show the signal changes with calibrators for a serum and a urine assay measured at intervals over 6–9 months using particle reagents stored in 5 mM glycine buffer with 0.1% azide as a bacteriostat. Both reagents were stable across these time periods suggesting a calibration stability in excess of 6 months.

5. Summary

Light scattering assays have been developed for large and small molecules and across the pM to mM concentration range. They are robust when carefully optimized, and due to their homogeneous nature are rapid and easily

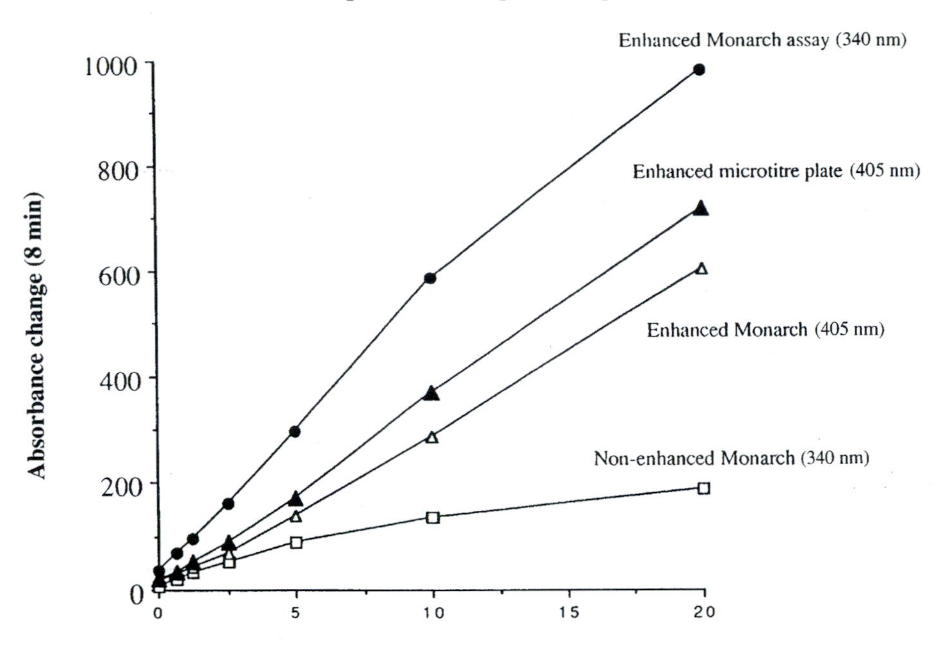

Figure 21. Performance of different assay formats, using a model of a serum β2m assay developed using 40-nm CMST particles. The assays were run according to the protocols in *Table 3*, and *Protocols 3* and *10*. The microtitre plate reader we used did not have a 340-nm filter and thus we have included a comparison of the signal change obtained on the centrifugal analyser at this wavelength. All formats could be used to provide an assay but only the enhanced method was applicable to the microtitre plate.

(preferably) automated. Whilst non-enhanced immunoturbidimetry is widely used particle enhancement has been viewed as less applicable to the non-specialist, this need no-longer be the case. Particle enhanced immunoassay are as robust and reliable as all other forms of immunoassay as long as careful attention is paid to assay optimization, which is again no different to any other form of immunoassay. The days of slide agglutination tests started long ago and although there are still several in existence, latex particles are now a technology with as good a record as any other immunoassay methodology.

References

1. Price, C. P., Spencer, K. and Whicher, J. (1983) Light-scattering immunoassay of specific proteins: a review. *Ann. Clin. Biochem.*, **20**, 1–14.
2. Newman, D. J., Hennebury, H. and Price, C. P. (1991) Particle enhanced immunoassay. *Ann. Clin. Biochem.* **29**, 22–42.

3. Griffin, C., Sutor, J. and Schull, B. (1994) *Microparticle reagent optimization.* Seradyn Inc., (available direct from the company).
4. Albert, W. H. and Staines, N. A. (ed.) (1993) Precipitation and agglutination methods. In *Methods in immunological analysis*, Vol. 1, pp. 134–213. VCH, Weinheim.
5. Price, C. P. and Newman, D. J. (1996) Light scattering immunoassay. In *Principles and practice of immunoassay*, 2nd Edn (ed. C. P. Price and D. J. Newman) pp. 443–80. Macmillan, London.
6. Heidelberger, M. and Kendall, F. W. (1935) Quantitative theory of the precipitin reaction; study of azoprotein-antibody system. *J. Exp. Med.*, **62**, 467–83.
7. Strutt, J. W. Rt. Hon. (Lord Rayleigh) (1871) On the scattering of light by small particles. *Phil. Mag.*, **41**, 447–54.
8. Cambiaso, C. L., Leek, A. E., De Steenwinkel, F., Billen, J. and Masson, P. L. (1977) Particle counting immunoassay (PACIA).I. A general method for the determination of antibodies, antigens and haptens. *J. Immunol. Methods*, **18**, 33–44.
9. De Mey, J. (1983) In *The preparation and use of gold probes in immunocytochemistry* (ed. J. M. Polak and S. Van Noorden) pp. 82–112. John Wright and Sons, Bristol.
10. Hellsing, K. (1978) Enhancing effects of non-ionic polymers on immunochemical reactions. In *Automated immunoanalysis* Part 1 (ed. R. Ritchie) pp. 1–44. Marcel Dekker, New York.
11. Singer, J. M. and Plotz, C. M. (1956) The latex fixation test. I. Application to the serologic diagnosis of rheumatoid arthritis. *Am. J. Med.*, **21**, 888–92.
12. Litchfield, W. J., Craig, A. R., Frey, W. A., Leflar, C. C., Looney, C. E. and Luddy, M. A. (1984) Novel shell/core particles for automated immunoassays. *Clin. Chem.*, **30**, 1489–93.
13. Kapmeyer, W. H., Pauly, H.-E. and Tuengler, P. (1988) Automated nephelometric immunoassays with novel shell/core particles. *J. Clin. Lab. Anal.*, **2**, 76–83.
14. Goding, J. W. (1978) Use of staphylococcal protein A as an immunological reagent. *J. Immunol. Methods*, **20**, 241–53.
15. Thakkar, H., Davey, C., Medcalf, E. A., Skingle, L., Craig, A. R., Newman, D. J. and Price, C. P. (1991) Stabilisation of turbidimetric immunoassay by covalent coupling of antibody to latex particles. *Clin. Chem.*, **37**, 1248–51.
16. Collet-Cassart, D., Limet, J. N., Van Krieken, L. and R. De Hertagh. (1989) Turbidimetric latex immunoassay of placental lactogen on microtitre plates. *Clin. Chem.*, **35**, 141–3.

Acknowledgements

This chapter could not have been written without the combined experience of many colleagues over the years particularly at St. Bartholomew's at The Royal London, Carol Davey, Peter Holownia, Helen Hennebury; at Dade Behring, Eileen Gorman, Alan Craig, John Thompson and Mike Largen.

6

Enzyme amplification: A means to develop fast ultrasensitive immunoassays

COLIN H. SELF, DAVID BATES and DAVID B. COOK

1. Introduction

Immunoassay has become a dominant tool in a wide variety of analytical areas. While simplicity and convenience for the user were very important aspects in this development, the demonstration that immunoassays of very high performance could be made was also critical. Enzyme amplification was developed to address both of these issues (1).

The fundamental problem of early immunoassays was that the enzyme labels were not able to generate sufficient signal simply by converting a single substrate into a product. Radio labels therefore remained dominant. However, much greater power of signal generation was required to allow more sensitive and faster assays, whether in immunometric systems or the inherently less sensitive competitive systems (2), which while slower as well as less sensitive, could still be improved (3). Enzyme amplification solved this problem without the need for elaborate and expensive instrumentation such as luminometers or fluorimeters.

The signal amplification of enzyme amplification has allowed extremely sensitive systems to be constructed. The high signal output also facilitates the convenience of very rapid assays where required. Simplicity stems from the fact that the systems lend themselves to colorimetric end points, read either visually or by means of simple colorimeter. Numerous demonstrations have shown that enzyme amplification provides a means of increasing the signal from an enzyme label in enzyme immunoassays by several orders of magnitude compared with conventional substrates. While colorimetric methodology is not normally considered capable of such high performance, when linked with enzyme amplification, colorimetric assays of extreme sensitivity can be constructed.

The increase in the signal generating capacity of the labelling enzyme is achieved by making it part of a system which gives rise to a product which is not itself detected, or indeed irreversibly used up in the signal generation, but instead acts as a catalytic activator for a secondary system which produces a much greater detectable response (1). The detectable signal is thus the product of two catalytic activities, the labelling enzyme and the activator.

The earliest approach (1) used a labelling enzyme that gave rise to an activator of a secondary system which was based on an activatable allosteric enzyme. Phosphofructokinase was used to produce fructose 1,6-biphosphate which in turn activated *E. coli* type 1 isoenzyme of pyruvate kinase when assayed at a limiting concentration of phosphoenolpyruvate substrate. This system exhibited a key feature of the principle that the activator is not consumed as a result of the activity of the secondary system.

A critical factor in further development of the principle was the realization that the activator could be consumed by the secondary system as long as it was reformed. This led to the development of cyclic amplifiers where the labelling enzyme produces a product such as a co-factor which is recycled by a cycling system which also produces the substance to be detected (1). In turn this led to the most widely used systems which employ alkaline phosphatase as the labelling enzyme and an NAD co-factor microcycle (4) as shown in *Figure 1*. In this system, alkaline phosphatase dephosphorylates NADP substrate to produce NAD. The NAD then becomes locked in a cycle. First it is reduced by highly NAD-specific alcohol dehydrogenase in the presence of ethanol to NADH. This is then oxidized to NAD by the enzyme diaphorase which concomitantly reduces INT-violet substrate to a highly coloured formazan. One molecule of formazan is produced in each 'turn' of the cycle. For given concentrations of the two enzymes the speed of cycling and thus formazan production is dependent upon the concentration of NAD produced from the activity of the alkaline phosphatase.

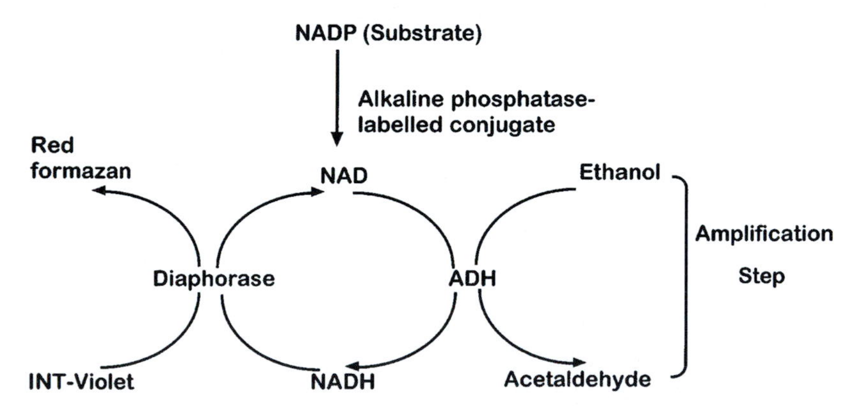

Figure 1. Principle of enzyme amplification for the detection of alkaline phosphatase label by the production of formazan.

Though the use of systems producing a coloured end-product are very convenient, cycles designed to generate fluorescent (5), thermometric (6) or electrochemical signals (7,8) have also been developed. The co-factor cycle is a redox process whereby the transfer of electrons from ethanol to INT-violet is catalysed. It was realized early in the development of the technology that the final chemical step could be by-passed and the process monitored amperometrically with an appropriate mediator and electrode. A system using hexacyanoferrate as electron acceptor which was reoxidized at an electrode surface was described by Stanley *et al.* (7). Subsequently Athey and McNeill (8) described an adaptation of the cycle where instead of diaphorase, NADH oxidase was used. The H_2O_2 generated was detected at an electrode set at 600 mV. The value of amperometric systems lies in their potential for the development of new applications of enzyme amplification, with important potential for even simpler systems for the user and the introduction of new formats such as, for example, systems allowing measurements to be made on whole blood or even allowing *in vivo* use.

The application of enzyme amplification to the quantification of alkaline phosphatase has resulted in systems capable of detecting less than 1 zeptomole (10^{-21} mole) of the enzyme both by colorimetric (9) and by fluorimetric determination (5). Among other applications of the general approach first described (1) have been examples in which FADP has been employed to produce FAD as the recycling component (10), and also one in which phosphoenolpyruvate is converted to pyruvate which is recycled in a system involving lactate oxidase and lactate dehydrogenase and detected thermometrically (6). All such systems have shown the expected benefits of enzyme amplification.

With regard to the high performance NAD-based cycling systems monitored by colorimetry, alkaline phosphatase has been found to be a particularly useful enzyme since it is capable of employing NADP or NADPH as substrates yielding NAD or NADH, respectively (11). This is important, as the cycle works equally well if NADH is produced as the triggering co-factor, rather than NAD. However, when using NADPH it is important that pig heart diaphorase is used. Microbiological diaphorase (which is commercially supplied extracted from *Clostridium kluyveri*) exhibits a much greater reaction with NADPH. Since the principle depends on the specificity of the diaphorase for NADH (as well as the alcohol dehydrogenase for NAD), if microbiological diaphorase is used NADP must be employed as the alkaline phosphatase substrate and not NADPH. This is also essential in the fluorescence cycle where the diaphorase substrate is resazurin which the pig heart enzyme does not use so efficiently.

Enzyme amplification has been used extensively in research and clinical applications, especially in the investigation of infectious diseases. Examples include the production of high performance kits for the determination of chlamydia (Dako Diagnostics Ltd), and hepatitis B and HIV I and II by Murex Diagnostics Ltd. In the latter case, the high sensitivity enables the use of unconcentrated urine as the sample thus avoiding the need for potentially hazardous blood sampling, as elegantly demonstrated by Connell *et al.* (12). The technique has also been applied to the assay of cytokines. Markham *et al.* have, for example, described a high performance assay for tumour necrosis factor with a detection limit of 1 pg/ml achieved by incorporating enzyme amplification into the assay system (13), and a wide variety of high sensitivity enzyme amplified ELISA kits are marketed by R & D Systems. The high sensitivity of the system has been also exploited in the diagnostic kit developed for urinary growth hormone by Novo Nordisk (14) and the recently launched kits for the estimation of total proinsulin and intact proinsulin from Dako Diagnostics Ltd. In addition, the technology has been applied to genetic testing. An early example was that of Coutlee *et al.* (15) who employed enzyme amplification for RNA determination. More recently, Minter (16) has employed biotinylated oligonucleotide probes hybridized to immobilized DNA with quantification with streptavidin–alkaline phosphatase conjugate and enzyme amplification.

Important methods employing enzyme amplification have also been reported for the measurement of anti-DNA antibodies (17), anti-glomerular basement membrane antibodies (18) and angiotensin converting enzyme employing a sandwich system (19). Another use of enzyme amplification has been to determine alkaline phosphatase itself as an important clinical marker. This has been achieved by means of 'immunoassisted' systems in which the enzyme is captured by a highly specific monoclonal antibody and then disclosed by the enzyme amplification system (4). Such a system was used to excellent effect in the diagnosis of hypophosphatasia from small first trimester chorionic villous samples (20).

Importantly, it has also been pointed out that an additional benefit of enzyme amplification can be an actual saving in cost when an amplified assay is compared to an otherwise identical but unamplified assay which requires more of an expensive antibody reagent (21).

2. Practical applications

The following discussion will be focused on applications in immunoassay but the fundamental considerations apply to the various other applications.

2.1 Water and buffers

This has been placed at the start of the practical section because of its importance. We have noted that the assays of a number of metabolites are extremely sensitive to the quality of water in which buffers are prepared, and consequently reliance cannot be placed on distillation, or even double distillation of water from certain sources. We have noted this to be a problem particularly with proinsulin and C-peptide (but interestingly not insulin), ACTH and PTH-related peptide (22). The reason resides chiefly in the immunochemical reaction but where ultrasensitive signalling systems are in use, water quality is of course particularly important.

Our work has shown that water drawn from a suitable deionizer, such as the Millipore Milli-Q, is suitable providing the treatment cartridges are not near to exhaustion. Water quality from such devices is often monitored only in terms of conductivity. However, it appears that the problems we have encountered relate to the Total Oxidizable Carbon (TOC) content of water, which conductivity measurements do not monitor. Use of water produced by cartridges near to the end of their life has frequently resulted in complete failure of binding without any indication of a problem from the conductivity meter on the deionizing system. In such cases, the meter usually indicates problems one or two weeks later.

We have observed that tap water in our laboratory in Newcastle upon Tyne exhibits considerable fluorescence which persists to some extent even after double distillation suggesting volatile organic materials to be the source of the problem. Such materials are presumably removed by charcoal filters. We have found that measuring the fluorescence of water sources gives a reliable guide to the efficacy of water, but for absolute reliability TOC as well as conductivity should be monitored at the production device. For these reasons we recommend the use of fresh Milli-Q water at all times.

2.2 Colorimetric enzyme amplification

The most common applications of enzyme amplification have been in ultrasensitive detection of sandwich ELISA assays employing high concentrations of ligand under conditions of reagent excess, such as in the ultrasensitive assay for TSH described in *Protocol 4*. The description relates to the use of reagents in kit form, but the same principles apply to any ELISA providing that proper care is taken in the design of the assay. The practical considerations noted below apply to many types of assay but are especially important with such highly sensitive assays as the enzyme amplification system allows. However, while care is required in the manipulative steps it is not more than would be expected from good laboratory practice.

1. The quality of the enzyme conjugated antibody is critical to optimal performance and minimization of non-specific binding. Heterobifunctional coupling chemistries (23) using, for example, SMCC and SPDP (Pierce and Warriner, Chester, UK) have proved to be superior to one-step methods using glutaraldehyde.
2. Conjugate quality is also improved by careful fractionation, assessment and choice of individual fractions. A suitable and convenient method is high pressure gel permeation chromatography on TSK-3000 or TSK-4000 columns (Anachem, UK).
3. The use of antibody fragments (Fab, Fab′ or $F(ab')_2$ is often preferable to whole antibodies. (see Chapter 1).

4. Careful optimization of conjugate buffers is essential. Factors to be considered include buffer type and pH, ionic strength, detergent concentration and protein concentration. The following buffer is suitable for storage and use of conjugates in amplified assays of alkaline phosphatase: 100 mM triethanolamine HCl (pH 7.5), 6% (w/v) BSA, 1 mM $MgCl_2$, 150 mM NaCl, 0.05% Triton X-100 together with 15 mM sodium azide as preservative.
5. Efficient washing is critical to the elimination of background and non-specific binding. Again buffer type, pH, ionic strength and detergent concentration may all need to be optimized separately.
6. The use of phosphate buffer in wash solutions immediately prior to incubation with alkaline phosphatase must be avoided as this will cause product inhibition of the enzyme and compromise the sensitivity achievable. Conjugate buffer itself should not contain phosphate; Tris buffers are recommended.
7. Great care should be taken to avoid cross-contamination of reagents. Use a separate pipette and reagent trough for conjugate. If storage of reagents for further use is required, ensure that all pipettes and containers that contact the containers are sterile.
8. As an alternative to conjugation of the signal antibody with alkaline phosphatase, biotin may be used. The biotin is subsequently detected with commercially available streptavidin-conjugated alkaline phosphatase (e.g. Sigma, Piece and Warriner, Vector) which gives lower background and non-specific signal than ordinary avidin. Deglycosylated forms of avidin are, however, available from Sigma (Extravidin) and Pierce and Warriner (Neutravidin) which may also be suitable. Vector Laboratories recommend their 'Avidin D'.

Protocol 1. The basic ELISA assay

1. Perform the ELISA assay in microtitre plates in the usual manner, using highly pure reagents. In stages where alkaline phosphatase conjugates are incubated, avoid the use of phosphate buffers; traces of phosphate may inhibit the enzyme. Use Tris buffer at pH 8. Phosphate buffers may be used when incubating the sample with capture antibody alone, followed by careful washing.
2. Perform all immunochemical and enzyme detection incubation steps with the plate enclosed in a suitable container to minimize temperature variation across the plate (24) to which ultrasensitive assays, particularly with enzyme labels, are prone. A small plastic box is convenient and highly effective. Some workers avoid using wells at the edges of plates to nullify the potential for such 'edge effects'.

3. At the various wash stages employ at least six wash changes for best results as regards sensitivity and precision. Some workers insist on eight wash changes to minimize non-specific signal. Either slap the plate between washes on absorbent paper to free it of traces of wash buffer or employ a good automatic plate washer.

Protocol 2. Preparation of enzyme amplification reagents

1. Dissolve alcohol dehydrogenase from yeast (EC 1.1.1.1), 70 mg in 7 ml of 20 mM sodium phosphate buffer, pH 7.2. Extensively dialyse against four to five changes of the same buffer at 4°C.
2. Resuspend pig heart diaphorase (NADH:dye oxidoreductase; EC 1.6.4.3), 10 mg in 5 ml of 50 mM Tris–HCl buffer, pH 8.0. Extensively dialyse against four to five changes of 20 mM sodium phosphate buffer, pH 7.2.

Before dialysis an inert carrier protein such as BSA may be added (to 10 g/litre) to improve stability. After dialysis the reagents should be centrifuged or filtered to produce clear solutions.

Prepared enzyme reagents should be stored at –20°C or below. Sodium azide (1 g/litre) may be added as preservative.

3. Prepare substrate diluent: 50 mM diethanolamine, pH 9.5, 1 mM $MgCl_2$, 0.7 M ethanol.
4. Dissolve NADPH in substrate diluent to a concentration of 0.1 mM.
   - The purest available grade of NADPH should be used, but even 'ultrapure' reagent may be contaminated with low concentrations (typically $<0.1\%$) of NAD(H) leading to a significant background. If necessary, NADPH can be repurified by conventional ion-exchange chromatography on a suitable matrix such as Q-Sepharose. For NADP, that described as NAD-free (Boehringer-Mannheim) is satisfactory for many purposes.
   - Prepared reagent may be stored for 1 week at 4°C or for longer periods at or below –20°C. Sodium azide (1 g/litre) may be added as preservative. Great care should be taken to avoid contact with phosphatase-containing solutions, including bacterially contaminated solutions.
5. Prepare amplifier diluent: 20 mM sodium phosphate, pH 7.2, 1 mM INT-violet, 0.05% (w/v) Triton X-100. Store at 4°C.
6. Prepare working amplifier solution by adding both alcohol dehydrogenase and diaphorase to amplifier diluent. A suitable strength reagent is produced by adding 50 μl of each enzyme to 900 μl of diluent. The activity of the amplifier can be adjusted by varying these volumes.

NOTE: Enzyme amplification reagents are now available commercially in two forms, AMPAK and AmpliQ. AMPAK provides the two reagents in a freeze-dried form with a diluent, whereas AmpliQ is a ready to use formulation with all components liquid. The improved stability of AmpliQ has been achieved by replacing yeast alcohol dehydrogenase in the amplifier with the homologous enzyme from the mesophile *Zymomonas mobilis*, which is much more stable.

The reagents in AmpliQ have also been reformulated into a one-step amplifier for simplicity of use. In this both the dephosphorylation of NADPH and the cycling of NAD/NADH occur simultaneously. In consequence, the kinetics of INT reduction are different; in the two-step system (AMPAK), colour development is linear in the second phase, whereas in the one-step system (AmpliQ), colour develops throughout the incubation at a quadratic rate. Both systems can be stopped by the addition of acid and the red formazan read at 492 nm.

Because the reagents are provided in ready to use dropper bottles, and because only one incubation step is required, additional protocol options are possible with AmpliQ, which makes the reagents more convenient to use. Because of diffusion of NADH from the plastic surface, the performance of AmpliQ is enhanced by moderate shaking of the assay throughout the one-step incubation. Reagents in the two products are not interchangeable.

Protocol 3. Use of enzyme amplification reagents

Equipment and reagents

- Enzyme amplification reagents, AMPAK and AmpliQ (Dako Ltd, 16 Manor Courtyard, Hughenden Avenue, High Wycombe, Bucks. HP13 5RE, UK)
- Single and multi-channel pipettes
- Sterile tips and reagent troughs
- Sterile water
- Absorbent paper
- Stopping solution, 0.46 mol/litre sulphuric acid
- Plate shaker
- Plate reader (492 nm)

Methods

In both cases, the immunological stages of the assay should be performed as already described and the assay carefully washed in an appropriate buffer (phosphate should be avoided). Amplification reagents can be used as follows:

A. AMPAK

1. Reconstitute the substrate and amplifier in their appropriate diluents.
2. Add 100 μl substrate to each well, in a timed sequence, from a single or multi-channel pipette.
3. Incubate for 10–40 min at room temperature.

4. Add 100 μl amplifier, in the same time sequence.
5. Incubate for 10–20 min at room temperature.
6. Stop the reaction by adding 100 μl stopping solution

B. AmpliQ

AmpliQ reagents can be added in one of three ways:

1. Add two drops of Amplifier A to each well (approx. 100 μl) from the dropper bottle.
2. Add two drops of Amplifier B.

or

1. Add 100 μl Amplifier A to each well from a single or multi-channel pipette.
2. Add 100 μl Amplifier B.

Note: it is not necessary to time the interval between the addition of these reagents if adding them separately. The timing of the incubation should begin with the addition of Amplifier B.

or

1. Premix the required amounts of Amplifiers A + B (100 μl each per well).
2. Immediately pipette 200 μl into each well.
3. Incubate for 10–60 min at room temperature on a plate shaker.
4. Stop the reaction by adding 100 μl (or 2 drops) stopping solution.

For all three systems the plate should be read at 492 nm, as soon as possible, but within 30 min of stopping.

Protocol 4. Enzyme-amplified ELISA for TSH (and a comparison with pNPP detection)

Equipment and reagents

- Novoclone TSH ELISA (DAKO Ltd, 16 Manor Courtyard, Hughenden Avenue, High Wycombe, Bucks HP13 5RE, UK)
- 5 mM pNPP in diethanolamine buffer (1 M, pH 9.8)
- Single and multi-channel pipettes
- Sterile pipette tips and reagent troughs
- Sterile water
- Absorbent paper
- Plate reader (405 and 492 nm wavelength filters)
- Stopping solution: 0.46 mol/litre sulphuric acid

Capture of TSH

Following the instructions provided with the kit:

1. Add 75 μl of anti-TSH/alkaline phosphatase into each well.

Protocol 4. *Continued*

2. Add 25 μl sample or calibrator to each well. Calibrators are provided in the range 0.1 to 500 mU/litre.
3. Incubate for 2 h at room temperature on a plate shaker.
4. Discard the contents of the plate and drain on absorbent paper.
5. Wash the plate three times with the wash buffer provided.
6. Drain the plate on absorbent paper, slapping hard to remove the last traces of solution.

Amplified-enzyme detection (AMPAK)

1. Add 100 μl of reconstituted substrate to each well.
2. Incubate for 15 min at room temperature.
3. Add 100 μl of amplifier to each well.
4. Incubate for 15 min at room temperature.
5. Add 100 μl of stopping solution to each well.
6. Ensure the contents of the wells are mixed and free of bubbles.
7. Read the plate at 492 nm within 30 min of stopping.

Detection with pNPP

1. Add 200 μl of pNPP to each well.
2. Incubate for 60 min at room temperature.
3. Stop the reaction by adding 100 μl of 1 M NaOH to each well.
4. Ensure the contents of the wells are mixed and free of bubbles.
5. Read the plate at 405 nm within 30 min of stopping.

Specimen results of a TSH ELISA run as described (and with different timed incubations) are shown in *Figure 2*, which illustrates the amplification that can be achieved in a conventional assay of normal duration. For equal detection periods of 60 min, the amplification factor (signal response) is about 250-fold which, given equivalent precision in the two systems, translates into a similar improvement in detection limit. A second property of the two systems is also evident; in the pNPP system, signal increases in proportion to time, whereas in the amplified system, with two linear kinetic phases, signal increases in proportion to the second power of time, giving an approximate fourfold increase in signal for a doubling of incubation (25).

3. Fluorimetric enzyme amplification

Fluorimetry is more sensitive than spectrophotometry by several orders of magnitude (26). However, the problem of non-specific binding prevents the

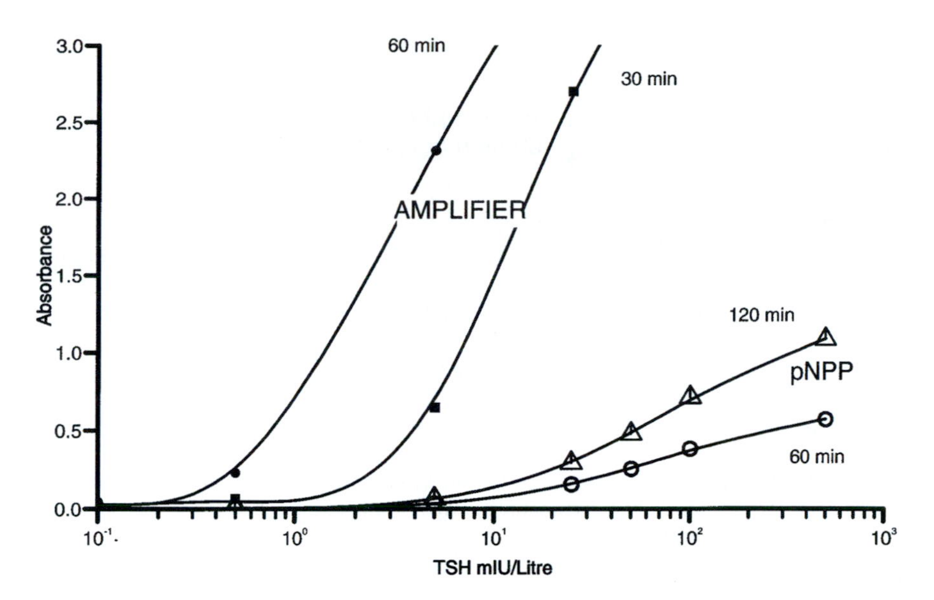

Figure 2. A comparison of enzyme amplified assay and pNPP in an immunoassay for TSH. For equal detection periods of 60 min, the amplification factor is 250-fold.

achievement of such impressive increase in sensitivity in immunoassay systems by simply substituting a fluorimetric step for a colorimetric determination (27). There are however fluorimetric alternatives for the quantification of the commonest enzymes used in immunoassays, including alkaline phosphatase which in addition to providing increase in sensitivity of about 5–10 fold have the additional advantage of extending the dynamic range.

As far as enzyme amplification is concerned, resazurin is a convenient substrate for diaphorase which can be used in place of INT-violet as shown in *Figure 3*. While in practice it is difficult to employ a fluorimetric approach to improve the excellent sensitivity provided by colorimetric enzyme amplification, an advantage to be taken into consideration in choosing a fluorimetric technique is extended dynamic range.

Resazurin was used in milk testing from the 1930s to detect contamination from dehydrogenase producing bacteria. In the 1960s, Guibault and Kramer (28) described general procedures for its application to the detection of dehydrogenases, whereby upon reduction the highly fluorescent pink product, resorufin, is generated. It is however far from an ideal substrate for analytical purposes since the blue colour of the resazurin substrate quenches the fluorescence of the resorufin product, limiting the sensitivity that might be considered achievable by the application of fluorescence. Thus a balance has to be chosen between the sensitivity and dynamic range of the assay. Sensitivity requires that the initial concentration of resazurin be limited to minimize the quenching, whilst such a strategy results in the exhaustion of the

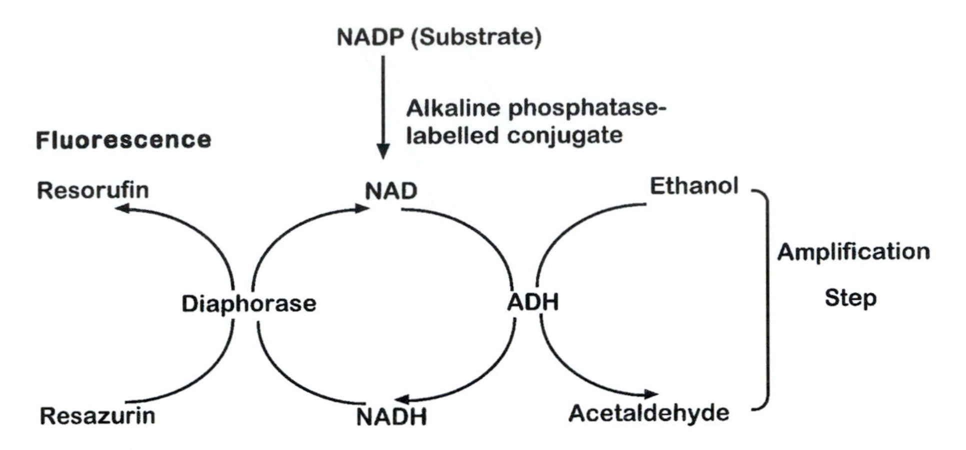

Figure 3. Application of fluorescence to the enzyme amplification principle. Resorufin is quantified fluorimetrically at an activation wavelength of about 560 nm and emission wavelength about 580 nm.

dye in a very short period with a resulting limitation on the dynamic range. Conversely when using a concentration of resazurin sufficiently high to improve the dynamic range, the sensitivity is limited by the inevitable quenching. Given these limitations, however, the authors have described an assay for proinsulin over a 500-fold range with a sensitivity of 0.017 pmol/litre (5).

When using fluorescence, the diaphorase must be from the microbiological source (*Clostridium kluyveri*) because resazurin is not a good substrate for the pig heart enzyme. As discussed above the residual reaction of the *Clostridium* enzyme with NADPH is much greater than for the pig heart (29), and it is therefore mandatory to use NADP and not NADPH as the triggering enzyme substrate. (Either NADP or NADPH may be used with pig heart diaphorase.) We have found that good results are derived with NADP ('NAD-free') crystallized monopotassium salt obtained from Boehringer Mannheim (reagent no. 1 179 969), without further purification.

There are a variety of commercial sources of suitable enzyme and we have used *Clostridium* diaphorase from Sigma, Biogenesis and Boehringer Mannheim. It remains essential to purify these enzymes from commercial sources by dialysis (25) to remove traces of NAD or NADH, which would otherwise result in considerable background produced by causing the cycle to operate without the introduction of the trigger by the action of alkaline phosphatase. The alcohol dehydrogenase is the same as for colorimetry since it is used with the same substrate, ethanol.

As with colorimetry it is necessary to determine the optimum balance of the concentrations of diaphorase and alcohol dehydrogenase in the mixture of amplification enzymes, and this must be checked with individual batches (25).

With fluorescence, there is an additional consideration imposed by the differing design of microplate fluorimeters available. Whereas most analysts would recommend the use of black plastic microtitre plates for the most sensitive fluorimetric estimations (and sometimes good results have been noted with opaque white plates intended for luminometry), some fluorimeters are designed so that the exciting light beam strikes the wells from below. Such designs obviously imply that transparent plates be used in which sensitivity may be compromised by light spillage from well to well.

Though an increasing number of microplate fluorimeters are now being introduced to the market, we have experience of three as described in the following protocol.

Protocol 5. Practical steps in enzyme amplification with fluorescence

The type of microtitre plate employed depends on the microplate fluorimeter available. For example:

Method

- Millipore Cytofluor—Transparent plates must be used since excitation is from below the wells.
- Life Sciences International Fluoroskan—Excitation is from above. Transparent plates may be used but better results are obtained with opaque black plates.
- Dynatech Fluorolite—Excitation from above. More sensitive detection of fluorescence in other applications has been observed with opaque black, and even better with opaque white plates intended for luminometry. (We have not used the Dynatech Fluorolite for resorufin fluorescence measurements.)

1. Prepare diethanolamine buffer; add 4.8 ml diethanolamine to 1 litre pure water. Bring to pH 9.5 with pure HCl.
2. Prepare substrate solution: 100 μM NADP in 50 mM diethanolamine buffer pH 9.5., containing 1 mM $MgCl_2$ and 1 μM $ZnSO_4$. It is convenient to keep stock solutions of 1 M $MgCl_2$ and 1 mM $ZnSO_4$ for this purpose when 10 μl of each can be added to 10 ml of the solution of NADP in buffer.
3. Prepare a fresh solution of 6.3 mg resazurin (Sigma) in 5 ml AR ethanol.
4. Prepare solution of amplifying mixture in 0.1 M phosphate buffer pH 7.4 as follows:
   - Diaphorase 7.5 units*
   - Alcohol dehydrogenase 302 units*
   - Alcoholic resazurin solution 580 μl (final concentration of resazurin *c.* 0.29 mM)
   - 0.1 M phosphate buffer pH 7.4 to 10 ml (at this stage inhibition of alkaline phosphatase immaterial)

Protocol 1. *Continued*

Make up the amount of solution required for the number of wells employed. For a 96 well plate about 12 ml should be prepared. Unlike INT-violet there is no need to include Triton X-100 detergent in this solution in order to keep the product soluble.

*The exact amounts will need to be checked for each batch of enzyme.

5. Perform ELISA assay as above. Finally wash wells six times with wash buffer.
6. Drain the plate on absorbent paper, slapping hard to remove traces of solution.
7. Add the same volume of substrate reagent as used for coating the plastic with capture antibody so that all the surface captured enzyme is detected. This will usually be 100–200 μl.
8. Incubate for 30 min at ambient temperature (within a box). This incubation time may be extended to 60 min or more for greater sensitivity.
9. Add 100 μl amplification solution to each well. Incubate at room temperature for development of fluorescence (this will be visible).
10. Make fluorescence readings from about 10 min onwards if necessary for 1–2 h. The fluorescence is stable for this period of time and we have found that even with extended periods of incubation blank readings do not increase unduly providing scrupulously pure reagents have been employed.

Read the fluorescence on a suitable microplate reader. The excitation and fluorescence spectra of resorufin overlap considerably (5) and this places demands on choice of filters for isolation of excitation and fluorescent wavelengths. The instruments of which we have experience contain filter sets which are suitable for quantification of resorufin fluorescence. These are:

- Fluoroskan: Filter set 3—excitation spectrum bandwidth 524–564 nm, fluorescence spectrum bandwidth 572–608 nm.
- Cytofluor: Filter set C—excitation spectrum bandwidth 505–555 nm, fluorescence spectrum bandwidth 555–625 nm. Sensitivity setting 1 or 2 is usually satisfactory.

An example of an enzyme amplified fluorescent assay for proinsulin is shown in *Figure 4*, which demonstrates the different signals achieved with 0.29 mM resazurin and 0.875 mM resazurin. At the lesser concentration of the dye, greater sensitivity is achieved, and greater fluorescence is demonstrated at lower analyte concentrations because of minimal resazurin quenching. On the

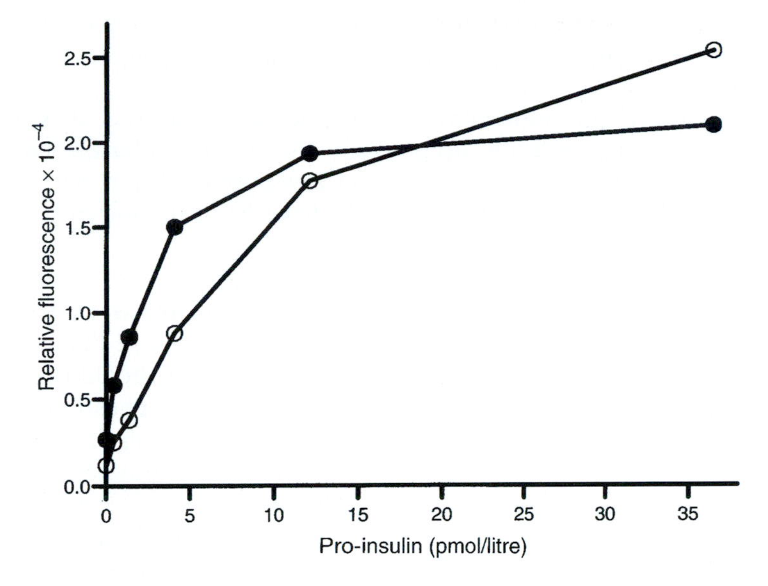

Figure 4. Fluorescence detection in enzyme amplified ELISA with resazurin substrate at 0.29 mM (●—●) and 0.875 mM (○—○). Sensitivity is greater with 0.29 mM dye which has been exhausted at 10 pmol/litre proinsulin. The greater dye concentration results in quenching at lower analyte concentration but produces more fluorescence at higher analyte levels as quenching is relieved when resazurin is consumed.

other hand, the greater concentration of resazurin gives rise to an extended dynamic range.

4. Conclusions

Amongst the various methods that have been described for high-performance non-isotopic immunoassays, enzyme amplification holds the particular attraction of not requiring specialist instrumentation. It may be performed on a microplate spectrophotometer that an immunodiagnostic or research laboratory is likely to possess almost as a matter of course. The preparation of reagents for enzyme amplification is certainly not particularly demanding of the staff of a typical analytical laboratory. However, for laboratories wishing to avoid preparation of reagents, high-performance kits are available for application to any assay system involving alkaline phosphatase (AmPak, AmpliQ, Dako Diagnostics Ltd). These are capable of providing a marked improvement in assay sensitivity compared with conventional colorimetric or fluorimetric procedures. The excellent results obtainable from applying enzyme amplification to existing assay systems has been underlined by Markham *et al.* (30). In their comparative study, which shows the degree to which enzyme amplification out-performs horseradish peroxidase label, they draw attention to the ease with which the system can be applied to existing assay systems stating that, 'The amplification system can significantly increase sensitivity of in-house ELISAs'.

Where biotinylated antibodies have previously been used, it is a simple matter to use or substitute streptavidin–alkaline phosphatase conjugates (which are available at high quality from a variety of commercial sources), even when alkaline phosphatase may not have been the enzyme originally used for avidin signalling.

For those wishing to employ the fluorimetric approach to enzyme amplification, the number of microplate fluorimeters available is increasing and such instrumentation is being acquired by an increasing range of laboratories.

Finally there is an aspect of enzyme amplification that should not be overlooked; the signal may be detectable visually. In some applications this may be all that is required. For example, in neonatal screening of TSH for the recognition of congenital hypothyroidism exact quantification of the signal is considered by some not to be necessary. The rare, but critical, positive signal (either coloured or fluorescent) from the many thousands of samples screened could readily be observed by eye in a microtitre plate indicating a patient requiring follow-up investigation without the need for instrumentation at all.

Appendix: List of suppliers

Anachem, Charles Street, Luton, Beds, LU2 0EB, UK

Biogenesis Ltd, 7 New Fields, Stinsford Road, Poole, BH17 0NF, UK

Boehringer–Mannheim GmbH, Postfach 31 01 20, Sandhofer Strasse 116, D-6800 Mannheim, Germany

DAKO Ltd, 16 Manor Courtyard, Hughenden Avenue, High Wycombe, Bucks. HP13 5RE, UK

Molecular Devices Corporation, 3180 Porter Drive, Palo Alto, CA 94304, USA

Murex Diagnostics Ltd, Central Road, Temple Hill, Dartford, Kent, DA1 5LR, UK

Novo Nordisk. Novo Allé, DK 2880 Bagsværd, Denmark

Nunc A/S, Postbox 280, DK-4000 Roskilde, Denmark

Pierce and Warriner (UK) Ltd, 44 Upper Northgate Street, Chester, CH1 4EF, UK

R & D Systems Europe, Ltd, 4–10 The Quadrant, Barton Lane, Abingdon, OX14 3YS

Sigma Chemical Co., Fancy Road, Poole, Dorset, BH17 7BR, UK

Vector Laboratories Inc., 30 Ingold Road, Burlingame, CA 94010, USA
3 Accent Park, Bakewell Road, Orton Southgate, Peterborough PE2 6XS, UK.

References

1. Self, C. H. (1981). European Patent Publication number 0027036.
2. Jackson, T. M. and Ekins, R. P. (1986). *J. Immunol. Met.* **87**, 13.
3. Stanley, C. J., Johannsson, A. and Self, C. H. *J. Immunol. Met.* **83**, 89.
4. Self, C. H. (1985). *J. Immunol. Met.* **76,** 389.
5. Cook, D. B. and Self, C. H. (1993). *Clin. Chem.* **39**, 965.
6. Mecklenburg, M., Lindbladh, C., Li, H., Mosbach, K. and Danielsson, B. (1993). *Analyt. Biochem.* **212,** 388.
7. Stanley, C. J., Cox, R. B., Cardosi, M. F. and Turner, A. P. F. (1988). *J. Immunol. Met.* **112**, 153.
8. Athey, D. and McNeill, C. J. (1994). *J. Immunol. Met.* **176**, 153.
9. Bates, D. L. (1995). *Intl. Labmate*, **20**, 11.
10. Obzhansky, D. M., Rabin, B. R., Simons, D. M., Tseng, S. Y., Severino, D. M., Eggelte, H. *et al.* (1991). *Clin. Chem.* **37**, 1513.
11. Morton, R. K. (1955). *Biochem. J.* **61**, 240.
12. Connell, J. A., Parry, J. V., Mortimer, P. P., Duncan, R. J. S. McLean, K. A., Johnson, A. M. *et al.* (1990). Lancet, **335**, 1366.
13. Markham, R., Young, L. and Fraser, I. S. (1996). *Eur. Cytokine Netw.* **6**, 4954.
14. Main, K. M., Jarden, M., Angelo, L., Dinesen, B., Hertel, N. T., Juul, A. *et al.* (1994). *J. Clin. Endocrinol. Metab.* **79**, 865.
15. Coutlee, F., Viscidi, R. P. and Yolken, R. H. (1989). *J. Clin. Microbiol.* **27**, 1002.
16. Minter, S. (1995). Novel Amplification Technologies Conference, Washington, DC, USA.
17. Jones, J. V., Mansour, M., Sadi, J. D. and Carr. R. I. (1989). *J. Immunol. Met.* **118**, 79.
18. Saxena, R., Isaksson, B., Bygren, P. and Wieslander, J. (1989). *J. Immunol. Met.* **118**, 73.
19. Stevens, J., Danilov, S., Fanburg, B. L. and Lanzillo, J. J. (1990). *J. Immunol. Met.* **132**, 263.
20. Warren, R. C., McKenzie, C. F., Rodeck, C. H., Moscoso, G., Brock, D. J. H. and Barron, L. (1985). *Lancet* **2**, 856.
21. Carr, R. I., Mansour, M., Sadi, D., James, H. and Jones, J. V. (1987). *J. Immunol. Met.* **98**, 201.
22. Dhahir, F. J. (1992). Ph.D Thesis, University of Newcastle upon Tyne.
23. Ishikawa, E., Imagawa, M., Hashida, S., Yoshitake, S., Hamaguchi, Y. and Ueno, T. (1983). *J. Immunoassay* **4**, 209.
24. Kemeny, D. M. (1991). In *A practical guide to ELISA*, p. 79. Pergamon Press, Oxford.
25. Johannsson, A. and Bates, D. L. (1988). In *ELISA and other solid phase immunoassays* (ed. D. M. Kemeny and S. J. Challacombe), pp. 85–106. John Wiley, Colchester, UK.
26. Guilbault, G. C. (1990). *Practical fluorescence* (2nd edn). Marcel Dekker, New York.
27. Porstmann, T. and Kiessig, S. T. (1992). *J. Immunol. Met.* **150**, 5.
28. Guilbault, G. C. and Kramer, D. M. (1965). *Anal. Chem.* **37**, 120.
29. Bergmeyer, H. U. (ed.) (1983). *Methods of enzymatic analysis* (3rd edn), Vol. II, pp 179–80. Verlag Chemie, Weinheim, Germany.
30. Markham, R., Young, L. and Fraser, I. S. (1996). *Aust. J. of Med. Sci.* **17**, 37.

7

Equipment and automation; Appendix on dose-response curve fitting

IAIN HOWES and RAYMOND EDWARDS

1. Introduction

The basic aspects of automation have been reviewed together with various degrees of automation and some practical applications. It has been impractical to consider all aspects of automation. Automation refers to tasks carried out by mechanical or electrical devices (essentially machines) assisting in the execution of part or all of the analytical technique.

For many immunodiagnostic applications, high throughput is an essential requirement. This can be greatly assisted by automation. Since immunoassays consist of a number of sequential steps which may be automated, this automation takes various forms ranging from the automation of a single step, illustrated by multiwell gamma counters, through multicomponent modular systems, to fully integrated totally automated systems of which immunoassay autoanalysers are working examples.

1.1 Features of automation

Automation can influence analytical techniques by:

- increasing the control of multiple variables such as pipetting and washing which taken together will significantly enhance assay performance
- removing operator variability leading to greater precision
- reducing cost per test by reducing or removing labour intensive operations
- removing repetitive tasks from the analyst
- increasing throughput of samples
- decreasing turnaround time
- provision of skill-independent analysis

- increasing maintenance costs within the laboratory—sophisticated automated equipment requires regular servicing
- reducing exposure to infectious samples
- reducing transcription errors

2. Steps in immunoassay analysis

The immunoassay process, from the receipt of the samples in the laboratory to the final release of the results, consists of a number of sequential steps which may be automated. These are:

- sample identification—bar-coding
- worklist generation
- transfer of sample from primary tube to reaction vessel
- addition of immunoreagents
- incubation
- separation where necessary
- signal generation
- signal detection
- curve fitting—response calibration
- quality control
- report generation
- export of results to LIMS

2.1 Sample identification—bar-coding

Unique bar codes applied to the tube from which the sample will be taken for analysis in the assay positively identify the samples and facilitate tracking of their progress through the immunoassay. Bar codes are also applied to the reagents. Prior to pipetting, the bar codes are read, this allows precise accounting of which sample and reagents have been added to each tube.

2.2 Worklist generation

The worklist will link together the samples, their bar-codes and the assays, i.e. it will precisely define which assays are to be performed on which samples.

2.3 Transfer of samples to reaction vessel

The appropriate volume of sample is aspirated from the machine sample tubes and dispensed into the machine reaction vessel. Some processors manipulate samples over large distances, i.e. from, for example, LP4 size (75

× 11 mm) machine sample tubes on one side of the equipment to, for example, microtitre plates on the other; whilst other machines transfer samples from a reservoir to a specific reaction cuvette in an integral assay specific cartridge.

2.4 Addition of immunoreagents

Once the sample aliquot has been dispensed into the appropriate reaction vessel the immunoreagents must be dispensed. The nature of these immunoreagents is dependent upon the format of the assay but will always include either labelled antibody or labelled antigen. If the solid phase component of the assay is not already present in the reaction vessel, this must also be added, e.g. antibody-coated beads.

2.5 Incubation

The sample and immunoreagents are allowed to react for a fixed amount of time. This reaction may proceed at ambient temperature or elevated temperature depending upon the format of the assay and the machine on which the analysis is being performed. Some assays may require agitation of the reaction vessel throughout the incubation time, many machines can support this requirement.

2.6 Separation where necessary

In homogeneous systems, the signal is read directly from the milieu in the reaction vessel. In heterogeneous systems, signal reading and subsequent quantitation of the tracer will depend upon the assay format. With radioactive tracers, after washing, the residual radioactivity can be measured directly in a gamma or beta counter. Assays employing enzymes as labels require a second incubation step in which the enzyme is reacted with its substrate prior to measurement, e.g. reaction of horseradish peroxidase (HRP) enzyme labels with tetramethylbenzidine (TMB) prior to spectrophotometric measurement at 450 nm.

2.7 Signal generation

Immunoassays employing enzymes as labels require incubation with the enzyme substrate in order to quantify the enzyme.

2.8 Signal detection

This step involves quantitation of the bound or unbound label. The detection system will vary according to the nature of the label employed.

2.9 Curve fitting—response calibration

The final result of a quantitative immunoassay procedure is the concentration of analyte in each sample. The technique itself generates the response for each sample, e.g. counts per minute (cpm) for radioactivity, or an absorbance (or optical density, OD) for colorimetric analysis. Calibrants of known analyte concentration are included in the assay. The curve fitting/response calibration step of the analysis uses a mathematical model to describe the way in which the response varies with the concentration. The concentration of analyte in the unknown samples is interpolated from the response using this model (1).

2.10 Quality control

The inclusion of quality control pools with established target values and ranges in assay runs will allow performance of the assays to be monitored. For the results of an assay to be released, the quality control criteria for that analyte must be satisfied.

2.11 Report generation

This is the final presentation of the results of the immunoassay analysis. The worklist and the assay results are linked together to give for each patient the concentration of each analyte measured. There are numerous formats of the final assay report dependent upon the mode in which the machine operates and in the sophistication of the software.

Analysers operating in batch mode will produce a list of patient names and assay results. Analysers operating in random access mode may present the information grouped so that all the assays performed on a patient appear together. For example:

(1) Analyser operating in batch mode

T4 results

F. Smith	142 nmol/litre
J. Brown	97 nmol/litre
A. Nother	123 nmol/litre

(2) Analyser operating in random access mode

F. Smith

Thyroid function results

T4	142 nmol/litre
T3	1.5 nmol/litre
TSH	1.8 mU/litre
free T4	21 pmol/litre

2.12 Export of results to LIMS

The results of the analyses are added to the laboratory information management system (LIMS). This will include updating of patient records and record keeping of appropriate quality control procedures. Systems with bi-directional interfaces may be linked directly to the LIMS system and the process carried out automatically.

3. Equipment

Automated equipment can be classified as follows according to the functions performed:

(1) automated equipment assisting manual operations, e.g. pipettors/diluters;
(2) modular automation which can automate a number of steps, such as sample/reagent processors and microtitre plate assay processors;
(3) robotic arms which may link together modular systems ultimately resulting in automated immunoassay systems;
(4) fully integrated automated equipment which can perform all the steps necessary for immunoassay analysis.

This classification represents a general framework. It will not always be possible to categorize equipment precisely.

The functions of various types of equipment are described in *Table 1*. A more extensive review of equipment can be found in reference 2.

3.1 Bar code readers

The variety of bar code equipment ranges from the simple hand-held 'wand' device, which is used to manually scan the bar codes, to machine-mounted readers, which scan the samples and reagents as they pass. The hand-held readers can be directly interfaced to worklist generators. The machine-mounted readers are already part of an integrated system.

3.2 Worklist generators

The worklist generating software available is capable of linking together samples and analyses required, the resulting program generated can be used to control equipment such as sample/reagent processors.

3.3 Pippettors/diluters

There are electronic pipettes available which, once programmed, will aspirate and dispense volumes of sample for analysis. More sophisticated equipment can aspirate a sample and dispense this together with reagent. They can also perform replicate dispensing from a single aspiration.

3.4 Reagent dispensers

There are a number of devices available which can dispense a programmed volume of reagent into tubes or microtitre wells. These range from hand-held

Table 1. Classification of equipment

	Individual equipment	**Signal detectors**	**Sample/reagent processors**	**Assay processors**	**Robotic linking arm**	**Integrated analyser**
Sample identification bar-coding	Bar code reader	No	Yes	No	N/A	Yes
Worklist generation	Stand alone software	No	Yes	No	N/A	Yes
Transfer of samples	Pipettes/diluters	No	Yes	No	Can link	Yes
Addition of immunoreagents	Reagent dispensers	No	Yes	Yes	Can Link	Yes
Incubation	Incubator/shakers	No	No	Yes	Can link	Yes
Separation	Washers	No	No	Yes	Can link	Yes
Signal generation	Reagent dispensers	No	No	Yes	Can link	Yes
Signal detection/quantitation	Signal detectors	Yes	No			
Curve fitting/response calibration	Stand-alone software	Yes	No	Yes	N/A	Yes
Quality control	Control charts and software	No	No	Yes	N/A	Yes
Report generation	N/A	No	No	Yes	N/A	Yes
Export of results to LIMS	N/A	No	No	Yes	N/A	Yes

Sample reagent processors and assay processors may be linked by robotic arms.

multichannel pipettes for adding reagents to a single plate, to dedicated machines capable of precisely adding reagent to a large number of plates very quickly.

3.5 Incubators/shakers

Equipment in which microtitre plates are incubated has options for control of temperature (often up to 45 °C) and the degree of agitation, as well as the incubation time.

3.6 Washers

There are numerous automatic plate washers available. The wash volume, soak time and number of wash cycles can usually be programmed by the operator. Other options such as the format of the wash heads (8, 12, 16, 24 or 96 way) and whether the strips are washed sequentially or individually is a function of the equipment. Some machines have the flexibility to adapt the washing to the profile of the wells.

3.7 Signal detectors

There are a range of detectors available to quantify radioactivity, absorbance (or optical density, OD), fluorescence and luminescence.

3.7.1 Radioactive counters

Radioactivity is measured in a gamma or beta counter yielding a response which is the number of radioactive decay events detected per unit time (frequently one minute).

The simplest situation is a single well counter into which the operator inserts each tube in turn, equipment is available which has automated this process.

Multiwell counters are available in which a number of tubes are counted simultaneously. The basic machines require the operator to load the tubes into trays and load the trays into the counter, more sophisticated equipment can do this automatically. The use of lead shielding in multiwell counters has led to minimization of cross-talk between the wells.

3.7.2 Spectrophotometers

The absorbance or optical density (OD) of a chromophore is measured in a spectrophotometer at the appropriate wavelength. Equipment ranges from spectrophotometers which can accommodate tubes and those which require cuvettes to multidetector microplate spectrophotometers.

There may be a single filter which can be changed or a number of filters available on the machine from which two can be selected simultaneously to facilitate bichromatic measurements. Most spectrophotometers give linear response $OD \leqslant 3.0$ and many have the capacity to measure $OD > 3.0$. Some equipment has the capability to perform kinetic measurements.

3.7.3 Fluorimeters

Fluorimeters measure either the intensity or the degree of polarization of the emitted fluorescence. More sophisticated instruments provide an initial excitation pulse with a delay before measurement of the emitted fluorescence. These instruments are known as time resolved fluorimeters. Both the duration of the initial excitation pulse and the time delay prior to measurement can be varied. Some fluorimeters can also perform kinetic measurements.

3.7.4 Luminometers

Luminometers quantify the light emitted from luminescent reactions. They are essentially very sensitive photomultiplier tubes. On some instruments the integration time may be varied.

3.8 Response curve calibration software

There are a number of software packages available which can calculate the response curve from the calibrators in the assay. Once this has been achieved, the unknown concentration of the samples can be determined from their response, by comparison with the response curve. Several mathematical models are available to provide the response curve. (see Appendix).

3.9 Quality control

A number of programs are available which plot QC data from the assays. A number employ rules suggested by Westgard (3) and may be used to gauge assay performance. Only results from assays showing satisfactory performance should be reported.

3.9.1 Implications for accreditation

When steps of immunoassay analysis are automated it is necessary to have the controlling software perform a number of system checks to confirm that the machine has completed these steps satisfactorily. Monitoring sensors on pumps, probes and reagent reservoirs record their status. This can ensure the security of the system and aid conformity to GLP.

3.10 Sample/reagent processors

Sample/reagent processors (see *Table 2*) identify samples via their bar code, aspirate them from the machine sample tube, place an aliquot of sample in the appropriate machine reaction vessel and add immunoreagents.

3.10.1 Probe options/disposable tips

The number of probes on an instrument will govern the speed at which aspiration and dispensing can be performed. Teflon-coated probes tend to have lower adsorption characteristics than stainless steel probes, and give fewer problems with sample carry over. The problem with stainless steel probes may be resolved by employing extensive washing routines, particularly of the outside of the probes.

Disposable tips will eliminate the possibility of carry over, but changing tips will inevitably slow the machine down.

3.10.2 Variable probe spacing

Machines with fixed probed spacing are restricted to using sample and reaction vessels which can be accessed with this spacing. Variable probe spacing will allow much more flexibility in terms of the assays and the containers which will fit on the machine.

3.10.3 Liquid level detection system

It is mandatory for sample/reagent processors to have some form of liquid level detection. This will ensure that the probe is in contact with the liquids whilst aspirating.

3.10.4 Processing speed

The throughput in terms of the number of assay tubes which can be set up per unit time is a function of the speed of the mechanical operation involved in the process, i.e. x, y, z movement of the probe and the stroke time of the pumps. The processing speed will fundamentally depend on the type of assay, volume required for analysis, and number of reagents to be added.

3.10.5 Tubes which can be used on the machine

This describes the available options of tubes from which the machine can aspirate and those which it can dispense into.

3.11 Assay processors

Assay processors (*Table 3*) can incubate the assay, wash the assay, add substrate for reaction with the enzyme label, read the assay, perform response calibration and generate a report.

3.11.1 Capacity

The capacity refers to how many plates may be loaded onto the machine for processing. If there is only one dispensing unit, washing unit and reading unit, the plates must be processed sequentially, although one plate may be read whilst another is being washed. Sophisticated software (time management software, TMS) is required to schedule the operations of each assay.

3.11.2 Dispensing unit

The dispensing unit is described in terms of how many reagents may be used and which volumes can be accommodated. There may also be the option for different modes of dispensing, i.e. through a single probe or multiprobe manifold.

Table 2. Sample/reagent processors

Equipment	Supplier	Probe options (± disposable tips option)	Variable probe spacing	Liquid level sensing	Processing speed	Tubes
Guardian 2000	Zenyx	4 dual nozzle teflon-coated stainless steel probes (–)	Yes (9–36 mm)	Conductance	Adds reagents to 96 wells in MTP in <3 min	7–16 mm primary tubes, various reagent bottles, MTP
Guardian SP I/II	Zenyx	2 dual or single nozzle teflon-coated stainless steel probes (+)	No	Conductance	Full stroke of piston pump i n 0.75 s	7–16 mm primary tubes, various reagent bottles, MTP
Mega	Tecan	4 teflon-coated stainless steel probes (–)	No probes set at 18 mm	Capacitance	XYZ movement <1 s	Various tube sizes and MTP
Megaflex	Tecan	4 teflon coated stainless steel probes (–)	Yes (9–36 mm)	Capacitance	XYZ movement <1 s	Various tubes sizes and MTP
Biomek 2000	Beckman	1–8 probes with interchangeable tools (+)	No	Sonic	Dependent on tools being used	Various tubes sizes and MTP
Multiprobe 101	Canberra Packard	1 stainless steel probe, can be upgraded to 4 (–) (+ if upgraded)	No (yes if upgraded)	Modified capacitance	300 samples per hour	10, 12, 13, 16, 17 mm primary tubes; 33, 60, 150 ml reagent troughs; MTP
Multiprobe 204	Canberra Packard	4 stainless steel probes (+)	Yes (9–31 mm)	Modified capacitance	1200 samples per hour. Adds samples to 96 well MTP in <5 min	10, 12, 13, 16, 17 mm primary tubes; 33, 60, 150 ml reagent troughs; MTP
Mark 5	DPC	4 teflon-coated stainless steel probes (–)	No probes set at 18 mm	Conductance	Up to 1000 samples may be transferred per hour	10–17 mm primary tubes, MTP
Genesis	Tecan	4 or 8 stainless steel probes (+)	Yes (up to 38 mm)	Capacitance	XYZ movement <1 s	Various size tubes, reagent containers, MTP

RSP 5031	Tecan	1 teflon-coated stainless steel probe (–)	No	Capacitance	XYZ movement <1 s	10–18 mm primary tubes, MTP
RSP 8051	Tecan	4 teflon coated stainless steel probes (+)	Yes (9–280 mm)	Capacitance	XYZ movement <1 s. Up to 2000 samples per hour.	Various size tubes and MTP
Microlab 2200	Hamilton	Up to 8 probes (+)	Yes (9 or 18 mm)	Capacitance	25–30 samples per hour	Various tube sizes and MTP
Microlab AT	Hamilton	12 probes (+)	No	Sensors identify change between air and sample	Transfer 96 samples and reagent to MTP <7 min	13–17 mm tubes, MTP
Quasar	Zenyx	4 stainless steel probes (–)	No	Capacitance	Transfer 96 samples to MTP <5 min. Transfer 96 samples and reagent to tubes <8 min	Various size tubes, MTP
Lissy	Zinsser Analytic	4 stainless steel probes, teflon coated (+)	Yes (18–38 mm)	Capacitance	Up to 1500 pipetting transfers per hour	Tube sizes >OD 8 mm <OD 18 mm, MTP. Up to 192 tubes and 6 plates or 400 tubes

NB. All sample/reagent processors above have accuracy ≤1% and precision %cv ≤1%.
MTP = microtitre plate.

Table 3. Assay processors (microtitre plate)

Equipment	Supplier	Capacity	Dispensing unit	Washing unit	Reader
ELISA Processor III	Behring	20 reagent positions, 2 wash solutions, 10 plates processed simultaneously	25–300 μl dispensing single dispense into each well	Up to 6 washes per cycle; up to 120 second soak time	Filters available: 405 nm, 450 nm, 492 nm, 570 nm, 650 nm, 340 nm. OD <2.7 linear
FAME	Hamilton	16 reagent positions, 3 wash solutions, 20 microtitre plates; may be expanded to 24 reagents, 6 wash solutions, 30 microtitre plates	25–300 μl dispensing single dispense into each well after addition a programmable mix is available	Programmable washes and soak times	Filters available 340–750 nm supplied with 405 nm, 340 nm, 492 nm, 450 nm, 620 nm.
DIAS	Dynex	8 reagents, 2 washers, 1–3 stackers each with up to 6 plates	10–500 μl dispensed. 6 or 8 reagents through eight-way manifolds. 2 reagents dispensed through single probes	Programmable up to 99 wash cycles and 9999 second soak time	Filters available in the range 340–850 nm OD <3.0 linear
OMNI	Bio-Tek	25 reagents, 3 wash buffers, 18 plates	20–200 μl dispensed from one of two probes	Programmable washes and soak time	400–750nm OD range 0–3.0

3.11.3 Washing unit

The characteristics of the washing unit are similar to those already discussed for modular washers.

3.11.4 Signal detection

The standard filters available and performance characteristics are described in *Table 3*.

3.12 Robotic arms

Microprocessor controlled robotic arms (*Table 4*) are capable of moving items around the worksurface and even outside the workstation. They can link systems together to give integrated automatic immunoassay analysers, e.g. microtitre plate assays prepared on a sample processor can be transferred into an adjacent assay processor.

A range of processors employ technology similar to that of the RoMa arms and are fully open systems capable of automating ELISA assays. Due to the totally open nature of the system, any primary tube may be used and there is a choice of labelling technologies.

3.13 Fully automated analysers

Fully automated analysers fall into two main groups:

1. *Open systems* (*Tables 5(a,b,c)*)permit the user to substitute some of their own reagents for those of the suppliers. In some circumstances the use of these reagents will expand the menu of assays available on the equipment.
2. *Closed systems* (*Tables 6(a,b,c)*) which restrict the user solely to reagents supplied by the manufacturer.

Table 4. Robotic arms—linking systems

Equipment	Supplier	Description
ORCA	Beckman	ORCA is an anthropomorphic-like arm mounted on a rail along which linear motion can be accomplished. The arm has five axes of motion, which provide a wide range of positional freedom, and a servo gripping mechanism.
TRAC	Tecan	TRAC is an anthropomorphic-like arm which is stationary mounted.
RoMa	Tecan	The RoMa consists of a pair of prongs which can act as grippers. The gripping unit can rotate around the central *Z* axis along which it can also move up and down. It is suspended from an arm which can also move in the *X* and *Y* directions, allowing access to any part of the workstation.
Sideloader	Beckman	The sideloader is a robotic arm which can rotate about a central *Z* axis, it can also move up and down along this axis and has the ability to extend its reach.

Table 5(a). Fully automated systems (open)

Equipment	**Supplier**	**Immunoassay technology**	**Signal technology**
Advantage	Nichols Diagnostics Ltd	Magnetic microparticles coated with antibody or antigen	Direct chemiluminescent label detected in luminometer. Dual system acridinium esters or isoluminol.
Amerlite Processing Centre	Ortho Diagnostics Ltd	Coated microtitre wells	HRP enzyme label with chemiluminescent substrate. Enhanced luminescence read in luminometer.
Autodelfia	Wallac	Coated microtitre wells	Lanthanide chelate label. Time resolved fluorescence measured.
Brio	Radim	Coated tube and coated microtitre wells	Enzyme label with colorimetric substrate.
Cobas Core	Roche	Coated beads	HRP enzyme label with TMB substrate. End point absorbance measurement in spectrophotometer.
ES300	Boehringer Mannheim	Coated tubes	HRP enzyme label with ABTS substrate.
Liamat S300	Byk Sangtek	Coated tube and coated bead	Chemiluminescent label detected in luminometer.
Riamat 280	Byk Sangtek	Coated tube and coated bead	Radioactive label measured by gamma scintillation counter.

HRP = Horseradish peroxidase.
TMB = Tetramethylbenzidine.
ABTS = 2,2′-azinodi(3-ethylbenzothiazoline-6-sulphonate).

Table 5(b).

Equipment	Sample holders	Volume required for analysis	Maximum number of specimens per run		Maximum number of analytes per run
			(1 analyte)	(3 analytes)	
Advantage	Variety of specimen tubes in racks 10 × 75 mm to 16 × 100 mm with continuous feed access up to 120 tubes	250 μl	120	120	15
Amerlite Processing Center	Rack for 96 primary or secondery tubes	210–300 μl	288	288	3
Autodelfia	Sample racks of 12 samples per rack	325–400 μl	432	368	8
Brio	60 tube racks	Assay dependent	120	120	9
Cobas Core	Racks for primary or secondary tubes	205–450 μl	120	90	5
ES300	Disposable 2 ml containers	160–350 μl	150	50	12
Liamat	Rack for 140 specimens in a variety of tubes	220–400 μl	280	100	4
Riamat 280	Rack for 140 specimens in a variety of containers	220–400 μl	280	100	4

Table 5(c).

Equipment	Tests per hour	Time to first result	Mode	Calibration	Analytes available
Advantage	90–170	37 min	R, C, S	Bar-coded master cure on reagents two-point calibration	
Amerlite Processing Center	About 150	40–71 min	R	4 calibrators every day; 6 for full calibration	T, F, Onc, TDM, Ferritin, Cortisol, CK-MB, Inf
Autodelfia	Up to 150	95–270 min	B	2 calibrators per run; recalibration with change of lot no. 6–7 calibrators	T, F, Onc, Ferritin, GH, Cortisol, β_2 microglobulin
Brio	Up to 384	Assay specific	R or B	5 or 6 calibrators, full standard curve every day	T, F, Onc, Ferritin, Inf, IgE, Anti Gliadin IgG/IgM
Cobas Core	Up to 150	30–120 min	R or B, C, S	2–6 calibrators, recalibration up to 6 weeks	T, F, Onc, Inf, Ferrritin, IgE, TORCH, Anti H. pylori
ES300	80	50–240 min	B	Single point recalibration every 14 days	T, F, Onc, Inf, Tropanin T, Ferritin, Cortisol, TDM
Liamat	100	2 h–overnight	B(sel), C	Up to 8 calibrators every 4 weeks	T, F, Onc, Digoxin, Ferritin, HVA, VMA, β_2 microglobulin
Riamat	100	Assay specific	B(sel), C	Test specific	T, F, Onc, Ferritin, C-peptide, Cortisol, Digoxin

B = batch; R = random access; B(sel) = selective batch; C = continuous access; S = STAT facility; T = thyroid function; F = fertility; Onc = tumour markers; Inf = infectious diseases.

3.13.1 Immunoassay technology

The immunoassay technology (*Tables 5(a)* and *6(a)*) defines the type of assay in terms of the nature of the antibody and type of tracer employed.

3.13.2 Signal technology

The signal is described in terms of how it is generated and how it is read (*Tables 5(a)* and *6(a)*).

3.13.3 Sample holders

The type of sample holders (*Tables 5(b)* and *6(b)*) compatible with the machine and the number of samples that may be loaded prior to a run may severely limit the usefulness of the machine. If primary sample tubes, e.g. vacutainers, cannot be loaded directly into the machine, the samples have to be transferred to the appropriate sample vessels. If decanting (transfer) is necessary, for sample identification to be maintained it will be necessary to apply new bar codes to the new vessels.

3.13.4 Volume required for analysis

Each machine has a 'dead volume' that is volume of sample which is unavailable for analysis. The volume required for analysis (*Tables 5(b)* and *6(b)*) is the sum of the volume required for assay analysis and the 'dead volume'. Where there are limited amounts of sample, e.g. from neonates, the volume required for analysis will restrict the number of analytes which can be measured on each sample.

3.13.5 Maximum number of specimens per run

This represents the maximum number of samples which can be loaded onto the machine prior to each run. On machines which can analyse more than one analyte the maximum number of samples will vary according with the number of analytes required for each sample.

3.13.6 Maximum number of analytes per run

The maximum number of analytes per run (*Tables 5(b)* and *6(b)*) is the maximum number of different analytes which can be measured on a sample in a single assay run.

3.13.7 Tests per hour

This reflects the throughput of the machine, but will be a function of both the type and number of assays performed in the assay run (*Tables 5(c)* and *6(c)*).

3.13.8 Time to first result

This is the time after initiating processing that the results become available (*Tables 5(c)* and *6(c)*).

Table 6(a). Fully automated systems (closed)

Equipment	Supplier	Immunoassay technology	Signal technology
Access	Sanofi	Magnetic particle EIA with AP-labelled tracer	Dioxethane based chemilumniscent substrate, chemiluminescence read in luminometer
ACS 180 Plus	Corning	Magnetic particle IA with acridinium ester labelled tracer	Oxidation of acridinium ester to produce chemiluminescence, read in luminometer
AIA–1200	Tosoh	Magnetic particle EIA with AP-labelled tracer	4MUP substrate, end-point fluorescence read in fluorimeter
AIA–600	Tosoh	Magnetic particle EIA with AP-labelled tracer	4MUP substrate, end-point fluorescence read in fluorimeter
Auraflex	Organon Teknica	Ceramic magnetizable particle EIA with AP-labelled tracer	4MUP substrate, fluorescence read in fluorimeter
Axsym	Abbott	Low molecular mass, homogeneous FPIA. Large molecular mass MEIA	Fluorescein label polarization read in fluorimeter for FPIA. 4MUP substrate for MEIA, end-point fluorescence read in fluorimeter
Elecsys 2010	Boehringer Mannheim	Streptavidin-coated magnetic microparticles as solid phase, biotinylated antibody or antigen, $Ru(bpy)_3^{2+}$ novel ruthenium label.	Electrically triggered and controlled light emission (electrochemiluminescence) read in luminometer.
Immulite	DPC	Coated bead EIA with AP-labelled tracer	Chemiluminescent substrate adamantyl dioxethane phosphate, luminescence read in luminometer
Immuno 1	Technicon	Magnetic particle EIA with AP-labelled tracer. Latex agglutination turbidimetry homogeneous assay	PMP substrate, end-point absorption read in spectrophotometer. Absorbance read in spectrophotometer
IMx	Abbott	Low molecular mass, homogeneous FPIA. Large molecular mass MEIA	Fluorescein label polarization read in fluorimeter for FPIA. 4MUP substrate for MEIA, end-point fluorescence read in fluorimeter

Kryptor	Cis	Homogeneous system with europium cryptate antigen or antibody conjugates and an energy accepting fluorophore conjugated to antibody	Amplified non-radiative transfer of energy from a fluorescent donor to a nearby acceptor. Time resolved bichromatic readings at 625 nm and 665 nm
Minividas	BioMerieux	Coated cartridge EIA with AP-labelled tracer	4MUP substrate, end-point fluorescence read in fluorimeter
Opus Magnum	Behring	Low molecular mass, multilayer dry film technology with fluorescent labelled tracer. High molecular mass, coated cartridge EIA with AP-labelled tracer	Rhodamine dye labelled antigen measured in fluorimeter for multilayer film. 4MUP substrate for EIA, end-point fluorescence read in fluorimeter
Radius	BioRad	Coated microwell EIA with HRP-labelled tracer	TMB substrate absorbance read in spectrophotometer
SRI	Serono	Magnetic particle EIA with AP-labelled tracer	PMP substrate, end-point fluorescence read in fluorimeter
Stratus II	Baxter	Glass fibre impregnated filter paper RPI with AP-labelled tracer	4MUP substrate, end-point fluorescence read in fluorimeter
Vidas	BioMerieux	Coated cartridge EIA with AP-labelled tracer	4MUP substrate, end-point fluorescence read in fluorimeter

EIA = enzymoimmunoassay; IA = immunoassay; RPI = radial partition immunoassay; FPIA = fluorescence polarization immunoassay; MEIA = microparticle capture enzyme immunoassay; HRP = horseradish peroxidase; AP = alkaline phosphatase; 4MUP = 4-methyl umbelliferyl phosphate; TMB = tetramethylbenzidine.

Table 6(b).

Equipment	Sample holders	Volume required for analysis	Maximum number of specimens per run (1 analyte)	Maximum number of specimens per run (3 analytes)	Maximum number of analytes per run
Access	Variety of tubes	110 μl	60	60	24
ACS 180 plus	Double ring for 60 primary tubes or sample cups	60–250 μl	60	60	13
AIA–1200	Ten racks holding primary tubes or sample cups	110–325 μl	100	100	21
AIA–600	Sample cups in a continuous belt	110–325 μl	40	10	10
Auraflex	Carousel for 80 specimens in primary tubes 75–100 mm high, 12–16 mm wide, or micricups	205–400 μl	80	80	20
Axsym	Carousel for 60 primary tubes or 90 sample cups	60–250 μl	60 or 90	60 or 90	20
Elecsys 2010	30 samples on disk or 75 on Hitachi rack	Unknown	30 or 75	30 or 75	15
Immulite	Special sample cups loaded into a continuous chain	60–125 μl	47	18	12
Immuno 1	Cuvettes in racks of 6	s = 2–50 μl d = 70–500 μl	78	78	22
IMx	20 cuvettes or 24 reaction cells	150–250 μl	20	N/A	1
Kryptor	Up to 50 primary tubes or sample cups in segments each containing up to 10 positions	Typically 100 μl	50	20	15
Minividas	Specimens added to assay cartridge	s = 20–200 μl d = unknown	12	4	6
Opus magnum	Racks of 12 tubes or sample holders	70–160 μl	100 tubes 160 sample cups	100 tubes 160 sample cups	8
Radius	Up to 168 primary or secondary tubes	260–300 μl	168	168	12
SRI	Samples held in test cartridges with reagents	140–290 μl	179	60	22
Stratus II	Triple welled cup units	200 μl	30	N/A	1
Vidas	Specimen added to assay cartridge	s = 20–200 μl d = unknown	30	10	15

s = sample volume; d = dead volume.

3.13.9 Modes

In terms of the functional classification proposed for equipment, sample/reagent processors can operate in batch, random batch or random access modes. Since there are no sampling steps involved in the processing carried out by assay processors, mode in this context has no meaning.

Fully integrated immunoassay analysers can operate in any of the following modes: batch, random batch, random access, continuous access or STAT. A brief description of these modes follows.

Batch mode

Samples are collected until there are sufficient to perform a run. All samples are analysed for the same analyte, i.e. only a single assay is performed on each sample. Until the last sample loaded has been transferred, extra samples may be added to the run on some processors, but this is not a true STAT capability.

Random batch mode

Samples are collected until there are sufficient to perform a run. More than one, but only a limited number of assays may be performed on each sample. The samples do not have to be grouped in order for the assays. Until the last sample loaded has been transferred, extra sample may be added to the run on some machines, but this is not a true STAT capability.

Random access mode

Samples are loaded into the machine in any order. Any assay from the menu of the machine may be run on any sample. Multiple assays may be performed on any sample. Until the last sample loaded has been transferred, extra samples may be added to the assay run on some machines, but this is not a true STAT capability.

Continuous access mode

Samples are added to the machine at any time without affecting the processing of ongoing assays. Any assay from the menu of the machine may be run on the samples. Multiple assays may be run on any sample. Continuous access machines which allow precedence for urgent samples possess a STAT capability.

STAT mode

Urgent (STAT) samples may be added to the machine at any time without affecting processing of ongoing assays. The STAT samples are then processed immediately. Any test from the menu of the machine may be run on STAT samples. Multiple assays may be performed on STAT samples.

Table 6(c).

Equipment	Test per hour	Time to first result	Mode	Calibration	Analytes available
Access	100	25–35 min	R or B, C, S	Recalibration every 28 days	T, F, Onc, TDM, Ferritin, Cortisol, IgE
ACS 180	180	15 min	R or B, C, S	Recalibration 1–28 days with 2 calibrators	T, F, Onc, TDM, IgE, CK-MB, Digoxin, Ferritin, B12, Folate
AIA–1200	120	65 min	R or B, C, S	Recalibration every 28 days with 2 calibrators	T, F, Onc, Ferritin, β_2 microglobulin, GH, Insulin, CK-MB
AIA–600	60	65 min	R, C	Recalibration every 28 days with 2 calibrators	T, F, Onc, Ferritin, β_2 microglobulin, GH, Insulin, CK-MB
Auraflex	72	17–65 min	R or B, C, S	Full standard curve calibration monthly	T, F, CK-MB, β_2 microglobulin
Axsym	Mean 80, max. 120	8–41 min	R or B, C, S	6 or 2 calibrators, guaranteed for 14 days	T, F, Onc, TDM, Ferritin, Inf
Elecsys 2010	88	9 or 18 min	R or B, C, S	Mastercurve for batch bar-coded on reagents stable for up to 3 months. Only 2 point adjustment required	T, F, Onc, TDM, Ferritin, Inf, IgE
Immulite	Up to 120	43–73 min	R or B, C, S	2 calibrators every 4 weeks	T, F, Onc, Ferritin
Immuno 1	Up to 120	5–88 min	R or B, C, S	6 point calibration curve stable for 60 days	T, F, Onc, Folate, B12, Cortisol, TDM, Inf
IMx	36	13–34 min	B	6 calibrators (FPIA) monthly. 1 calibrator (MEIA) 2 weeks	T, F, Onc, Ferritin, CK-MB, B12, Insulin, Folate, Gly Hb, Inf, TDM
Kryptor			R or B, C, S	Bar-coded factory standardization curves for each lot of reagents, recalibration using one or two calibrators in duplicate	T, F, Onc, β_2 microglobulin

Minividas	Up to 24	20–150 min	R or B	Manufacturer's standard curve, 1 point recalibration every 14 days	T, F, Onc, Ferritin, IgE, TDM, Cortisol, β_2 microglobulin
Opus Magnum	Up to 190	6–18 min	R or B, C, S	3–6 calibrators every 6–8 weeks	T, F, Onc, Ferritin, CK-MB, Myoglobin, Inf, TDM
Radius	150	90–156 min	R or B	4 calibrators every 14 or 28 days. 1 drift control between calibrations	T, F, B12, Folate, Ferritin, DHEAS, Cortisol
SRI	Up to 60	39–80 min	R or B, C, S	6–8 calibrators when cartridge lot expires	T, F, Onc, Ferritin, Cortisol, DHEAS, Digoxin
Stratus II	49	8–90 min	B	Up to 6 calibrators guaranteed for 14 days	T, F, Onc, Ferritin, B12, Folate, IgE, Cortisol, TDM
Vidas	Up to 60	20–150 min	R or B	Manufacturer's standard curve, 1 point recalibration every 14 days	T, F, Onc, Ferritin, IgE, TDM, Cortisol, β_2 microglobulin

B = batch; R = random access; B(sel) = selective batch; C = continuous access; S = STAT facility; T = thyroid function; F = fertility; Onc = tumour markers; Inf = infectious diseases.

3.13.10 Calibration

The nature and frequency of calibration will vary from machine to machine. There may be a full standard curve with up to eight calibrators required with each assay run, or a single calibrator every 6–8 weeks. Some machines have self-contained assay cartridges on which there is factory determined calibration data bar-coded for direct input to the controlling microprocessor.

3.13.11 Analytes available

Tables 5(c) and *6(c)* list the range of assays currently available on each machine.

3.14 Clinical chemistry analysers

A limited number of immunoassays are becoming available on dedicated clinical chemistry analysers. In some cases, the equipment has been modified to include immunoassays; in other cases, the immunoassay technology has been adapted to run on the instrument.

3.15 Cleaning and disinfection

To ensue proper performance all laboratory equipment needs to be kept clean and regularly maintained. Automated equipment is no exception.

Cleaning methods include flushing the machine with purified water followed by alcohol or a range of solutions containing acid, detergent or digestive enzymes to remove proteinaceous deposits. Disinfection is either carried out using a weak hypochlorite solution or a propriety virocide. General user maintenance will include keeping the surfaces of the machine clean, checking tubing and seals for leaks and may extend to light oiling of some moving parts. Several manufacturers supply their own cleaning and disinfecting solutions together with recommended protocols.

A general protocol is given below.

1. Prior to each assay run prime the equipment with purified water and check the probes and tubing for leaks.
2. After each assay run prime the equipment with purified water. If the machine is not in use immediately after this, decontaminate the instrument:
   (a) Fill the probes and tubing with disinfectant and leave to stand for 10 min.
   (b) Prime the machine with at least three cycles of purified water.
   (c) Again fill the probes and tubing with disinfectant and leave to stand for 10 min.
   (d) Prime the machine with a further three cycles of purified water.
   (e) The machine should then be left with water in the tubing.

3.16 Safety features

There are a number of safety features available. These include motion sensors on the arms containing the probes which can avoid physical injury—robotic arms frequently move at speeds of 4 m/s!

Liquid waste containment and disposal, and safe transfer of disposable tips to sealable waste containers will reduce exposure to infectious agents.

4. Analytical considerations

The analytical performance of assays on automated equipment is evaluated in the same way as the analytical performance of any immunoassay reagents (4,5). The reagents are tested in their working environment and resultant errors will encompass any contribution from the mechanical processes involved in performing the assay. The criteria used are described below:

- within assay precision
- between assay precision
- matrix effects
- recovery
- parallelism
- cross-reaction
- heterophilic antibodies
- method comparison
- carry over
- performance on EQAS schemes (proficiency testing)
- sensitivity

The application of some of these criteria is described in Chapter 8.

5. Conclusions

Traditionally, immunoassays have been labour intensive procedures. The increased use of automated equipment and the integration of automated systems has led to a reduction of the dependency of the technique on skilled labour.

Automated immunoassay analysers represent a large capital investment and invoke considerable running costs both in term of reagents and maintenance. These costs can, however, be offset by increased throughput, decreased turnaround time and significantly reduced labour costs. In some cases, the increase in performance, particularly good precision, has led to immunoassays which were previously always performed in duplicate being performed in singleton. This will further reduce the unit cost of the assay.

The introduction of automated equipment will significantly alter work patterns within a laboratory. To ensure that the full potential is reached, appropriate training of staff is essential, as is an attempt to alleviate some of the 'technophobia' and maintain staff motivation.

The range of automated equipment available and under development is huge. Much attention is now focused on the development of fully automated immunoassay analysers, with continuous random access rapidly becoming a prerequisite.

The trend of automated immunoassay development has largely shadowed the diversification of the immunoassay technique. Initially developed and applied to the field of endocrinology, immunoassays are now commonplace in oncology, haematology and many other disciplines. The development of large menu automated analysers ultimately offering a fully automated pathology service is likely to continue. There is also considerable interest in the smaller, limited menu machines which operate from cartridges containing all reagents necessary for a particular assay. The use of these instruments in smaller hospital laboratories is bringing diagnosis closer to the patient. It is most likely that this trend will continue.

The type of 'microdot' technology already under development by Ekins, in which several analytes are determined simultaneously, will undoubtedly have a role to play in the future.

The choice of automation is complex and the final decision as to which equipment to use will depend as much on the type of laboratory as the performance of the automation.

References

1. Howes, I. (1996). In Immunoassays essential data. BIOS. Huddersfield, UK.
2. Wheeler, M. J. (ed.) (1995). The immunoassay kit directory: Series A Clinical chemistry, Vol. 3, Part 5 Equipment (December). Kluwer Academic Publishers. Lancaster, UK.
3. Westgard, J. O., Barry, P. L., Hunt, M. R. and Groth, T. (1981). Clin. Chem., **27**, 493–501.
4. Chan, D. W. (ed.) (1992). Immunoassay automation: A practical guide. Academic, San Diego, CA.
5. D'Souza, A. *et al.* (1994). MDD Evaluation Report MDD/94/03. Medical Devices Agency (Crown Copyright), Surbiton, UK.

Appendix

Data processing has become an important aspect of immunoassays, and is now an integral component of much equipment.

The process of converting detection signal (response) to dose (analytical concentration) is usually accomplished by the use of calibrants to construct a dose–response curve (standard curve). The dose–response curves generated by immunoassay are frequently complex in shape and construction of the dose–response curve (1) is often performed by onboard microprocessors. These microprocessors can also carry out other aspects of data processing such as:

- quality control
- sensitivity estimation
- precision profile calculation

Most dose–response curves are best simulated by a particular algorithm, the most appropriate for each assay type usually being found by trial and error.

The most commonly encountered and useful algorithms are listed below, and their advantages and disadvantages are given in *Table 1A*.

1. Linear interpolation

Adjacent points of the response calibration curve are joined with sections of straight line. The equation of the straight line is:

$$y = a + bx$$

where y = response; a = intercept on y-axis; b = gradient (slope) of the line; x = calibrant concentration.

Table 1A.

Model	Advantages	Disadvantages
1.	Simple. Will fit all data.	Not very robust, outliers may force model to deviate sharply. Results may be ambiguous. Only considers adjacent points. Dependent on number and spacing of calibrants.
2.	Simple. Will fit all data.	As above.
3.	Reasonably simple. Includes all calibrants. Reasonably robust. Statistical analysis possible.	The method is still dependent on the number and spacing of the standards.
4.	Reasonably simple. Includes all calibrants. Statistical analysis possible.	Not applicable to all situations, deviations from linearity occur at the exteremes of the concentration range.
5.	Robust. Includes all calibrants. Widely applicable. Statistical analysis possible.	May not fit all situations, in particular IRMA curves may present difficulties especially at the low concentration end.
6.	Robust. Includes all calibrants. Statistical analysis possible.	This is a theoretical equation and it may be difficult to obtain accurate values for some of the parameters.

2. Spline interpolation

Adjacent points of the response calibration curve are joined with sections of polynomial functions. The polynomial functions have equations:

$$y = a + bx + cx^2 + dx^3$$

where y = response; a, b, c and d are parameters specific to each section of curve; x = calibrant concentration.

3. Smoothed spline interpolation

Mathematical 'smoothing' which prevents changes in the sign of the gradient between adjacent points is applied to the curve constructed with splines, and can yield a smooth monotonic function.

$$y = F(x)$$

where y = response; F = appropriate resultant function; x = calibrant concentration.

4. LOGIT log

The LOGIT transformation

$$\text{LOGIT } Y = \log_e \frac{Y}{100 - Y}$$

where the most commonly used response Y is given by

$$Y = \frac{\text{(bound response)} - \text{(non-specific response)}}{\text{(maximum bound response)} - \text{(non-specific response)}}$$

is frequently applied to immunoassay data. A plot of LOGIT Y versus log concentration will often yield a straight line.

5. Logistic equations

The four parameter logistic equation is defined by

$$y = \frac{(a - d)}{1 + (x/c)^b} + d$$

where y = response; a = response at high asymptote (zero dose (RIA) or high dose (IRMA)); b = slope factor; $c = ED_{50}$, the concentration corresponding to 50% specific binding; d = response at low asymptote (high dose (RIA) or zero dose (IRMA)); x = calibrant concentration.

6. Law of mass action models

The single binding site equation based on the law of mass action proposed by Ekins has the form

$$[(D + p^* + c).b + q].R^2 + [c.(b + 1) + q - p^* - D)].R + c = 0$$

where R = free/bound ratio; p^* = tracer concentration; D = analyte dose; q = antibody concentration; c = 1/affinity constant; b = non-specific binding.

8

Quality assurance in immunodiagnostics

DAVID G. BULLOCK and FINLAY MACKENZIE

1. Introduction

1.1 The need for quality

The primary function of immunodiagnostic investigations, as with any health-care laboratory investigation, is to provide the clinician with analytical data obtained from examination of specimens submitted from patients and to assist in interpreting these results, in order to assist in diagnosis and control of therapy for the individual patient, for research or for public health purposes. Quality assurance is a global term to describe those means of ensuring that the results being issued from a laboratory are dependable and sufficiently accurate and precise to allow decisions to be taken with confidence, i.e. that the service is fit for the purpose. If such confidence is lacking, additional costs will accrue, not only in terms of inappropriate resource utilization through repeat testing etc., but also suffering of individual patients and, in some cases, the entire community. QA measures are not a luxury or optional extra, but must be an integral part of the investigational process in providing an appropriate and effective clinical service.

1.2 The purpose of quality assurance

Quality assurance improves test reliability through helping to minimize the variability arising from biological or analytical sources, which is inherent in all quantitative measurements or qualitative examinations. Overall, quality assurance seeks to guarantee 'the right result at the right time for the right investigation on the right specimen from the right patient, with result interpretation based on correct reference data'; to this may be added 'at the right price'. The introduction and continuation of an effective quality assurance programme, including appropriate quality assurance measures and schemes for internal quality control (IQC) and external quality assessment (EQA), is therefore essential in relation to preventive, diagnostic, curative and rehabilitative medicine. Quality assurance, IQC and EQA are required in all laboratories, to facilitate optimal patient care and to avoid the adverse consequences of quality failure.

1.3 Implementation of a quality assurance programme

Quality assurance in laboratory medicine is the means by which the reliability, precision and accuracy of investigations used in the support of optimal patient care can be achieved. Errors in analytical results may be related to various factors including the quality and education of laboratory staff, quality of reagents, apparatus and specimens, and the suitability of the techniques in use. Quality assurance seeks to minimize any variability in test results arising from these sources, and forms part of a quality system for the laboratory. Numerous books and articles have been published on the subject (e.g. 1–7). Though in many persons' perception quality assurance is synonymous with 'statistics' and therefore is simply avoided, statistics in fact merely provide a tool used within analytical quality control, which is only one aspect of QA, and the essence of quality assurance is the application of common sense to immunodiagnostic investigations.

2. Quality systems

There are several components of a quality system, which should all be in place and operating before the end product of a good quality laboratory service is likely to be fully achieved. Excessive attention to any of the individual components with neglect of others will not achieve lasting improvements in quality.

2.1 Quality assurance (QA)

QA is the total process whereby the quality of laboratory reports can be guaranteed. Quality assurance comprises all the various different measures taken to ensure reliability of investigations. These measures start with the selection of appropriate tests, continue with the obtaining of a satisfactory sample from the right patient, followed by accurate and precise analysis, prompt and correct recording of the result, with an appropriate interpretation and subsequent action on the result (4). Adequate documentation forms the basis of a quality assurance system in achieving standardization of methods and traceability of results on individual specimens. Regular monitoring of equipment, preventive maintenance and repair when needed are other important components of quality assurance.

Quality assurance must not be limited to the technical procedures performed in the laboratory. All those who send specimens to any clinical laboratory can contribute significantly to the reliability of the results through correct specimen collection and handling. A major source of variation can be linked with inappropriate specimen collection or subsequent treatment (e.g. by evaporation or delay in transport to the laboratory). Such variation must be minimized through careful training and adequate supervision, and the laboratory has a duty to ensure this.

2.2 Internal quality control (IQC)

IQC assesses, in real time, whether the performance of an individual laboratory or a field testing site is sufficiently similar to their previous performance for results to be used. Thus IQC controls laboratory reproducibility (measured as imprecision), enhancing the credibility of the laboratory in ensuring that sequential results are comparable and maintaining continuity of patient care. Most IQC procedures employ analysis of one or more defined control materials, and ascertain if the results obtained were within the limits of acceptability established previously within each laboratory.

2.3 External quality assessment (EQA)

EQA, by contrast with IQC, compares the performance of different testing sites. This is made possible by the analysis of an identical specimen at many laboratories, followed by the comparison of individual results with those of other sites and the 'correct' answer. The process is necessarily retrospective, and provides an assessment of performance rather than a true control for each test performed on patients' specimens.

2.4 Audit

Audit is a process of critical review of the laboratory. Internal audit is review of laboratory processes conducted by senior laboratory staff. Such reviews are aimed at measuring various parameters of performance, such as timeliness, accuracy and costs of reports, and identifying weak points in the system where errors can occur. External audit widens the input by involving others in the evaluation of laboratory services. The users of laboratory services (usually clinical colleagues) are asked how they perceive the quality and relevance of the service provided. Comparison of methods, working practices, costs and workload between laboratories by regular discussions between staff of laboratories forms a part of external audit. External audit should be a cyclic process of setting standards, examining compliance, and re-examining the standards (9,10).

2.5 Accreditation of laboratories

Accreditation is a process including inspection of laboratories by a third party to ensure conformity to pre-defined criteria. Factors which may be considered are numbers and qualifications of staff, facilities available, procedures used, evidence of quality assurance and quality control procedures, participation in external quality assessment schemes, adequate documentation, reporting procedures, safety, communications within the laboratory and with users, and management structure. Accreditation is commonly based initially on written statements of conformity (or reasons for any non-conformity) by laboratories, followed up by inspection to verify. Accreditation may be linked to a formal system of licensing, whereby only accredited laboratories are legally entitled to practice (or to receive payment for their services), or may be a voluntary system.

Within the UK, the most appropriate accreditation system for healthcare laboratories is that administered by Clinical Pathology Accreditation (UK) Ltd, an organization owned by the professional societies in laboratory medicine (11). Applicant laboratories claim compliance with the objective standards for laboratory facilities and function set by CPA, with compliance being checked by inspection as well as examination of the application. Compliance must be asserted annually, with inspection at 3–4 year intervals.

2.6 Validation of results

Validation is an attempt to measure quality by re-examination of specimens. This may be done by formal referral procedures, where results obtained with specimens submitted to reference laboratories are checked against the sending laboratory's results. Another approach, particularly in screening, is to select fixed percentages of negative and positive specimens reported by a testing laboratory and re-examine them in a reference laboratory. This procedure resembles the practice in IQC of reanalysing specimens from a previous analytical batch, and is of greatest importance where the assay is done in only a few laboratories.

2.7 Good manufacturing practice (GMP)

GMP is the system by which manufacturers of reagents and equipment ensure the quality of their products. The component factors of GMP include traceability of components and processes, documentation, quality control of components and quality control of product independent of the manufacturing procedures, adequate facilities, conformity with safety regulations, and proper labelling, packaging and product information.

GMP is relevant to clinical laboratories in two ways. Firstly, where alternatives exist, reagents and equipment should be purchased from manufacturers who can demonstrate that they follow GMP. Secondly, it must be recognized that some laboratories act as manufacturers through production of reagents, control materials or quality assessment materials. Such laboratories need to consider the application of GMP to their manufacturing activities.

2.8 Training and education

This represents probably the single most important component in a quality assurance programme. Issues to be addressed include the relevance of policies and curricula for the training of pathologists, scientists and technicians, both during primary training and in post. Professional status and career development are related factors. Immunodiagnostics must be included wherever appropriate, to ensure that the special features of these investigations are fully recognized by those carrying out or responsible for them.

Training of medical students and nurses in the appropriate and effective use of laboratory facilities is also important. Training needs should be continually monitored, and courses and workshops may be introduced where new needs arise (such as, for example, introduction of new methods or technologies), or where quality assessment programmes reveal the need for improvement.

2.9 Evaluation of reagents and equipment

Though many laboratories use EQA information as a 'buyers' guide', evaluation for suitability for use within a testing situation may make significant contributions to overall quality. Choice of equipment and reagents requires considerable thought and coordination, and may be cost-effective in reducing duplication of effort and preventing repetition of expensive mistakes. Reagents and equipment should be selected according to rational criteria which might include cost of purchase, revenue consequences, the appropriateness of the technology, robustness under local operating conditions, level of skill required to operate, and availability of spares and repair/support services. Unfortunately, initial evaluation cannot guarantee the quality of subsequent batches of reagents and some system of continual monitoring is often needed. Factors to be considered in selecting automated immunoassay systems have been reviewed in Chapter 7 and by Wheeler (12).

3. Quality assurance

Quality assurance (QA) (which is often equated with 'good laboratory practice') encompasses all measures taken to ensure reliability of investigations, starting from test selection, through obtaining a satisfactory sample from the right patient, analysing it and recording the result promptly and correctly, to appropriate interpretation and action on the result, with all procedures being documented for reference (1,4,8).

General quality assurance measures are essential in providing a secure basis within which analytical quality control can be effective. In addition to the aspects mentioned above, any testing site must employ well-chosen, reliable methods and equipment, carried out by trained, competent, motivated staff on correctly collected specimens, in an environment which is safe, clean, well-lit and appropriate to the task, and provide results which are recorded correctly. The immunodiagnostic laboratory provides a service to patients, and the turnaround of requests must be appropriately rapid.

Quality assurance measures, excluding analytical quality control (which deals with control of the analytical process itself, and is described in Sections 4 and 5 below), may be classified as:

- pre-analytical
- post-analytical
- non-analytical

3.1 Test selection and siting

In any situation where investigations are required, it is essential to examine critically the circumstances and needs, then agree between clinician and laboratory which should be carried out locally and which by a more distant laboratory. This should ensure that all investigations are relevant and reliable, with appropriate resourcing to assure quality. An increasing number of laboratory tests (mainly quantitative biochemical tests, but also including immunodiagnostic procedures) are now being considered for use in peripheral sites, using 'desktop' analysers or single-use devices designed for use in primary care. Although these are apparently simple to use, the quality of results does *not* come automatically with the equipment (13).

3.2 Pre-analytical quality assurance

3.2.1 Specimen collection and transport

The quality of the specimen dictates the maximum quality of the results obtained. It is not possible to get useful results on an incorrectly collected or deteriorated specimen, however good the analytical quality (though it is of course possible to get poor results on a reliable specimen). The laboratory thus has a responsibility to ensure that staff collecting specimens are well trained and do so reliably, and to ensure that the means for transporting specimens to the laboratory do not result in specimen degradation.

In collection, the appropriate anticoagulant (if any) must be used, and precautions taken against contamination. Special procedures may be necessary in some cases (e.g. cooling and prompt separation of serum for ACTH).

3.2.2 Specimen identification

Reliable identification of specimens, at all stages from the patient onwards, is essential to providing a good quality service. The effects on patients of transposition errors can often be much greater (and potentially fatal) than the effect of analytical failures in accuracy or precision. Particular attention must be paid to the unequivocal identification (usually by labelling, since positioning may not be fully reliable) of all intermediate containers, e.g. the tube to contain serum separated from clotted blood, the reaction tube in an analysis, in addition to the original specimen container. Bar-code labelling of specimens and/or request forms is increasingly used, and can facilitate reliable identification.

Appropriate clinical details to facilitate interpretation and assessment of plausibility should accompany the request. This may include stage of gestation or menstrual cycle in addition to age and sex; posture may be important, e.g. for renin and aldosterone assay.

3.3 Post-analytical quality assurance

3.3.1 Report identification and plausibility

Reliable identification of reports at all stages is essential to providing a good quality service. As for specimen identification, the effects of transposition or transcription errors can often be severe. Care must be exercised in all observation, recording, calculation and report transcription stages to ensure that the data are correct and attributed correctly. In calculation stages, it can be helpful to assess whether the result obtained is plausible, and when reporting to compare the investigation result with the clinical details on the request form and with the results of any other current or previous investigations on that patient.

3.3.2 Reference data and interpretation

Results should be reported with appropriate reference data. These may simply be the reference interval ('normal range') for the investigation, preferably taking thc patient's age, sex, etc. into account where appropriate, or the range expected in relevant disease states. A comment or interpretation is highly desirable where the clinician may not be fully conversant with interpretation, e.g. for a specialized immunodiagnostic investigation.

3.4 Non-analytical quality assurance

3.4.1 Laboratory facilities

Laboratory facilities should be adequate for the work to be carried out. Equipment, glassware, and other requirements should be available, and kept clean, well-maintained and in good working order. The exact requirements depend on the level of laboratory and the test repertoire, and, for example, for a multiwell gamma counter the well-to-well crossover and efficiency should be monitored. It is essential that sufficient spare parts for equipment are available, with expertise in maintenance, to permit full and appropriate use of all equipment.

3.4.2 Laboratory management and documentation

Management must be good, i.e. all resources (personnel, equipment, facilities, raw materials, such as blood, and finance) are used appropriately and effectively to the patients' benefit. Procedures used must not only be reliable but also be documented fully; documentation applies equally to quality control measures and managerial procedures as to the analytical methods themselves.

3.4.3 Staff selection, training and motivation

Staff are probably the laboratory's greatest asset (or liability). Staff should be selected for their aptitude, motivation and potential, given appropriate training (including retraining when required), and motivated continually. Only in this way will the laboratory be able to provide a service to patient care which is not only professional but is also seen to be professional; this includes a contribution to clinical interpretation of the investigations carried out.

3.5 Summary

Measures in quality assurance mostly comprise good common sense and seek to apply sound scientific and management principles to laboratory operation. These measures are in general relatively inexpensive and require no specialized equipment or tools, and there is thus no justification for not applying good quality assurance as a necessary basis for the more specialized techniques of analytical quality control.

4. Internal quality control

Internal quality control (IQC) assesses, in real time, whether the performance of an individual laboratory or testing site is sufficiently similar to their own previous performance for results to be used (8). It controls reproducibility (or precision), and facilitates continuity of patient care over time. Most IQC procedures employ analysis of a control material and compare the result with pre-set limits of acceptability—unsatisfactory sets of results may thereby be suppressed.

Many systems used in peripheral level testing are factory-calibrated (or require infrequent recalibration), and much of the variability of results originates from variations in operator technique. The analysis of an appropriate control material (it must be noted that not all materials are suitable for all systems—and may produce differing results even if they are suitable—as they are essentially artificial, stabilized materials) before starting analysis of a set of specimens can provide reassurance that the system and operator are working correctly. It is essential, however, that the results obtained are recorded, compared with acceptance limits and appropriate action taken: if results are outside limits, then the situation requires investigation and patients' specimens should *not* be analysed. The necessary procedures must be documented, and graphical presentations are helpful. A major source of variation can be specimen collection (see Section 3.2.1 above), but it is *not* possible to use IQC measures to control this, and this variation must be minimized through careful training and supervision procedures.

Automated analysers usually include IQC software and recording, but this should be reviewed for completeness and appropriateness to the laboratory's needs.

4.1 Simple IQC procedures

Some IQC procedures can be simple and inexpensive to implement. These include:

(1) recording of lot numbers of all reagents, calibrants and controls used, with particular attention when reagent and/or calibrant lots change;

(2) recording and monitoring of assay properties (e.g. non-specific binding, counts for zero standard), as a check on drift;

(3) recording and monitoring of instrument readings (e.g. absorbance for the calibration material), as a check on reagent and instrumental drift;

(4) analysis of one or more specimens (at differing concentrations) from the previous analytical batch, as a check on assay stability (provided that the specimens are stable for this period).

These procedures do not require sophisticated statistical techniques or expensive materials, and any laboratory should be capable of introducing these simple steps, which can provide invaluable information on assay performance.

4.2 Statistical quality control techniques

Most IQC procedures rely on introducing control specimens into each batch of analyses. These specimens must be stable and of reproducible composition (see Section 6), and the results obtained on them should then reflect the assay's performance with patients' specimens. Graphical and statistical analyses can be applied to the results to confirm whether or not the analytical process is 'in control' and thus whether or not the patients' results can be reported. It is important that the same material is not used for both calibration and IQC, or the IQC cannot be effective (1,2).

Classically, two specimens, one with normal and one with abnormal analyte concentrations, are analysed in each batch. The results are recorded, and compared with acceptance limits. These limits are derived from initial analysis of the control specimen in 20 batches, leading to calculation of the mean ('assigned' or expected value) and the standard deviation (SD). Subsequent results may conveniently be plotted on a control chart, with concentration as y axis and batch number (usually day) as x axis; this approach facilitates decision making if horizontal lines are drawn at the mean (expected) value and at 2 SD and 3 SD above and below the mean value. The SD used in these techniques should be at least as small as that in the laboratory; 'acceptance ranges' quoted with commercial materials are usually much too wide, as they include a substantial contribution from between-laboratory variation.

If the analysis is in control, results will be scattered randomly above and below the expected value, the distribution being such that only 1 in 20 (5%) will be more than 2 SD from the mean and only 1 in 100 (1%) more than 3 SD from the mean. Loss of precision, however, will yield a wider scatter of results, whereas loss of accuracy will cause a shift to higher or lower values; these changes can be seen on the graphical presentation ('control chart') and corrective action taken. The rules are:

Result >2 SD from mean	Warning—investigate method to avoid future problems
Result >3 SD from mean	Action—reject batch and investigate problem before repeating analyses

4.3 Validated control techniques

More complex and effective control rules have been formulated, notably by Westgard and co-workers, who have validated their power (to reject unsatisfactory batches and accept satisfactory batches) using computer simulation studies.

The so-called 'Westgard rules' (14) for the interpretation of control data using two materials analysed once in each batch have been published as a proposed selected method, and are given below. If both control results are within 2 SD from their target, the batch is accepted. If at least one control result is more than 2 SD from the target, the remaining rules are evaluated in turn, and the batch rejected if any one rule is satisfied. If none is satisfied, the batch is accepted, but the situation should be investigated before the next batch is analysed.

1_{2S}	Warning	One result more than 2 SD from target
1_{3S}	Action	One result more than 3 SD from target
2_{2S}	Action	Two consecutive results more than 2 SD from target in same direction
R_{4S}	Action	Difference between the two control results exceeds 4 SD
4_{1S}	Action	Four consecutive results more than 1 SD from target in same direction
$10_{\bar{x}}$	Action	Ten consecutive results same side of mean

In immunodiagnostic investigations, the use of at least two control materials at differing concentrations is strongly recommended. However, if there is minimal concentration-dependence and only one control material is analysed in each batch, a simplified modification known as 'Wheeler rules' (15) applies, with more stringent action rules.

4.4 Techniques using patients' data

For laboratories with a larger workload, and with access to calculation facilities, control techniques using patients' results can be helpful. These do not use QC materials, and are thus cheaper and not susceptible to errors due to matrix effects (1).

In essence, the mean of all patients' results for the day is calculated and used as a control parameter. Results which are very high or low must be excluded (though using exclusion limits which are too narrow makes the technique insensitive); application to analytes where results cover several orders of magnitude present problems, and log-transformation may be appropriate. Corrections may also be needed for the proportion of ambulant and hospitalized patients for proteins and for protein-bound hormones. Though the technique is helpful, it should not be used as the sole form of IQC.

4.5 Assay validation

In addition to the routine procedures described above, immunodiagnostic assays possess other characteristics which must be assessed on introduction of a new investigation (3,5,7). This assessment should be repeated periodically at intervals appropriate to the assay, and some aspects may also be addressed in EQA schemes (see Section 5). These important characteristics include:

- precision profiles and working range
- baseline security (specificity) and detection limit
- interference
- recovery of added analyte
- linearity
- parallelism
- antigen excess ('high dose hook' effect)

4.5.1 Precision profile

The precision profile for an immunodiagnostic investigation is an important and useful characteristic (3,16). In most assays, errors (expressed as CV) show a U-shaped relationship to concentration. The required information may be derived from multiple analysis of individual specimens across the concentration range, yielding CVs which can be plotted against each specimen's concentration. *Figure 1* shows an example precision profile generated in this way from replicate analysis of 70 specimens for total thyroxine. This technique can also be used with EQA data to generate an analogous 'between-laboratory agreement profile'.

More usually, however, a precision profile is obtained from the results of duplicate (or repeated) assay of clinical specimens. Here the concentration range is divided into bands ('bins'), and the paired results for all specimens

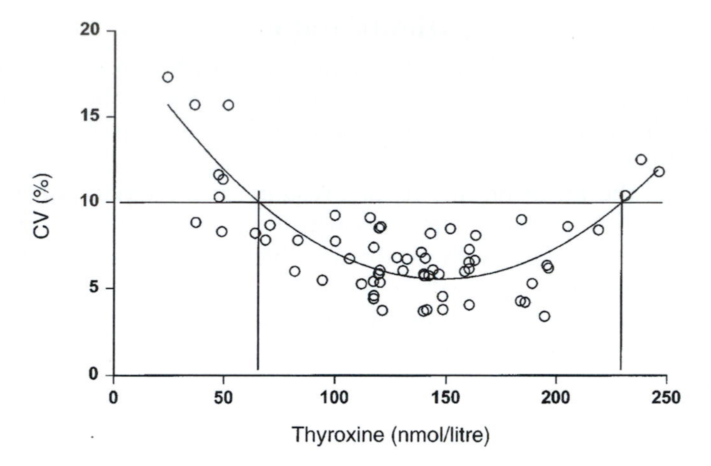

Figure 1. Precision profile for total thyroxine. Points represent the precision of replicate assay for 70 specimens; working range is 66–229 nmol/litre, defined as CV < 10%.

within each concentration band are pooled to give an imprecision estimate for the band. Only one CV point is thus available for each band, so many bands are desirable to give a better plot. Confidence in each point will be less, however, if only a few specimens contribute, e.g. at high concentrations.

The information is useful in three main applications:

(1) in assessing the assay's capability, and determining the working range;
(2) in providing a performance baseline, against which current performance can be compared if problems are encountered;
(3) in on-going monitoring of assay performance.

4.5.2 Working range

The working range of an assay is the range of concentrations over which it is usable in practice. This may be defined in precision terms, e.g. 'between-day CV < 10%', and determined from a precision profile. Results obtained which are above this range must be repeated after dilution, or reported as less than the lower limit of the range. For example, in *Figure 1* the working range (defined by a CV below 10%) is 66–229 nmol/litre. However, a working range based on precision alone is not sufficient; the upper limit of the working range must be checked to ensure that linearity still holds, and where relevant for 'high dose hook' effects (see Section 4.5.9 below).

4.5.3 Detection limit

The detection limit is often referred to as assay 'sensitivity'; this term should properly be reserved for clinical sensitivity (i.e. positivity in disease), though 'functional sensitivity' is now being used to clarify the application (17). The detection limit quantifies confidence in determining analyte presence from background variability, and is usually defined as 3 SD above the background signal. Commercial manufacturers frequently make exaggerated claims for this property, but using multiple replicate assay. Though replicate assay may be used to determine whether immunoreactive analyte is present or absent, such estimates are completely irrelevant to day-to-day use of the assay since assays are normally done singly or at most in duplicate. What is of practical importance is the lower limit of the working range, and all lower concentrations should be reported as less than this limit.

4.5.4 Baseline security

This can be assessed by challenging the assay with analyte-free specimen. Ideally this should be obtained by physiological means (e.g. suppression of thyrotropin to unmeasurable levels by administration of tri-iodothyronine (T3)), since selective depletion can alter the matrix and cause erroneous results. Assays should produce results below the detection limit, and such challenges have been used in EQA to confirm that laboratories are using appropriate detection limits for their assay (18).

4.5.5 Interference

Though antibody specificities are usually high in reliable assays, cross-reaction by related compounds may occur. This can be assessed by assaying specimens with and without the interferent to be tested, and any enhancement in the apparent analyte concentration expressed as a percentage cross-reaction:

Measured concentration = 27 μmol/litre

Apparent concentration with 50 μmol/litre interferent = 30 μmol/litre

Cross-reaction = 6% (i.e. $(30 - 27) \times 100/50$)

Potential interferents should be added at the concentration likely to be observed in clinical specimens. This is important, because apparent antiserum non-specificity may be insignificant if the cross-reactant is only present at very low concentration, whereas minor cross-reactivity becomes important with high concentration cross-reactants. For example, 10% cross-reaction by T3 in a thyroxine (T4) assay is clinically insignificant since a high level of 5 nmol/litre T3 would only yield an apparent additional 0.5 nmol/litre in a T4 of 100 nmol/litre. On the other hand, even 0.1% cross-reactivity by T4 could be important in yielding an apparent additional 0.1 nmol/litre T3.

Assays may also be subject to other interferences which are not related to the analyte (e.g. drugs, turbidity), and these should also be assessed where relevant. Any assay with a photometric, fluorescent or luminescent measuring principle may be subject to absorption or quenching effects from bilirubin, turbidity and other specimen properties.

4.5.6 Recovery

Assays should show quantitative recovery. This is studied by adding known increments of pure analyte to analyte-free (see Section 4.5.4) or low concentration specimens; care must be exercised in the design to avoid changing the matrix by dilution or by introducing interferents. After plotting the concentrations observed against the known additions, regression analysis should be used. In most cases, the endogenous analyte content will be unknown, and will be estimated by the intercept of the regression equation. Interpretation may be complicated by the lack of pure analyte or the need to add a different form (e.g. pure steroids in ethanol have been claimed to show divergence of behaviour from endogenous protein-bound steroids).

4.5.7 Linearity

Assay linearity can be assessed by making dilutions of a high concentration specimen to cover at least the working range. Clinical specimens with endogenous analyte should be used, though spiked specimens may be used if these are unavailable. The specimen should be diluted with low concentration (or analyte-free) serum, *not* with water or buffer, to ensure that the matrix is similar across the concentration range studied. As with recovery studies, graphical presentations and regression analysis provide the best approach to handling the data. Recovery and linearity studies should show similar assay performance, and any differences may be attributable to the use of endogenous analyte in linearity but exogenous analyte in recovery studies.

4.5.8 Parallelism

The demonstration of parallel (i.e. superimposable) dose response plots (of measured response against log dilution) on doubling dilutions of clinical specimens and of calibration standard indicates that the same analyte is being measured. Divergence of the plots indicates that the analyte in the standard is different, or in a different form, from that in clinical specimens; inability to superimpose the dose response plots for clinical specimens indicates the presence of different isoforms or different mixtures of isoforms.

4.5.9 Antigen excess

In some immunodiagnostic assays a recursive calibration curve may be seen, and, for example, specimens with high tumour marker or monoclonal immunoglobulin concentrations may yield apparently normal results. For investigations such as these where very high concentrations may be encountered it is therefore important to study recovery or linearity well above the working range, and modify the working range accordingly if necessary. The 'maximum secure concentration' is the concentration giving the same response as the maximum possible concentration on the recursive part of the calibration curve. Responses above this may arise from either of two concentrations, one below and the other above the maximum response, and so cannot be used.

Precautions which may be applied to extend the working range include the routine assay of specimens neat and in dilution (e.g. for tumour markers). For immunochemical protein assays, such as immunoglobulins, many modern automated systems include antigen excess checks, by monitoring reaction kinetics and/or by adding more specimen or standard to the reaction.

4.6 Clinical validation

In addition to analytical validation of investigations, clinical validation should be investigated wherever possible (3,5). This will give an invaluable insight into the usefulness of results in clinical practice. Aspects which should be assessed include:

1. Reference intervals ('normal range'); are there differences with age, with sex, or with other factors; can literature values or generally accepted ranges be used?
2. Cut-off values, used for example to classify results as 'normal' or 'abnormal', or as decision points in screening; can generally accepted action limits be used?
3. Clinical sensitivity ('positivity in disease'), representing the investigation's capability to yield a positive result in diseased patients.
4. Clinical specificity ('negativity in health'), the capability to yield a normal result in the absence of disease.
5. Predictive value of positive and negative tests. This requires knowledge of disease prevalence in the population to be investigated, and is particularly important for optimizing cut-offs in screening investigations.

4.7 Corrective action

Where IQC dictates batch rejection, or warns of impending trouble, the cause must be sought and procedures changed where needed to prevent recurrence (learning from errors is an important element of QA). The corrective action to be taken depends upon the type of problem, i.e. whether the assay has become imprecise or inaccurate (1,2,5,7).

Possible causes of loss of precision include:

- pipetting equipment problems (e.g. sampling probes, pump tubing)
- specimen problems (e.g. clots, gel serum separators)
- variable reaction/incubation timing
- variable reaction/incubation temperature
- instrument (e.g. photometer, counter) instability

Possible causes of loss of accuracy include:

- incorrect sample or reagent volumes pipetted
- incorrect reaction/incubation timing or temperature
- incorrect instrument setting (e.g. wavelength, energy)
- calculation errors
- deterioration of calibration material
- contamination

4.8 Summary

Simple but effective IQC procedures are available, and should be used universally. Without effective IQC programmes, laboratories cannot be sure that the results they produce are reliable, and they may thereby hinder rather than assist reliable healthcare.

5. External quality assessment

External quality assessment (EQA) addresses differences *between* testing sites, so there can be continuity of care over geography. Common mechanisms include the analysis of identical specimens at many laboratories, and the comparison of results with those of other sites and a 'correct' answer; the process is necessarily retrospective (8).

5.1 The purpose of EQA

Though in most situations EQA is considered to deal primarily with the assessment of individual laboratories, it in fact provides assessment of:

- the overall standard of performance (state of the art)
- the influence of analytical procedures (methods, reagents, instruments, calibration)
- individual laboratory performance
- the specimens distributed in the EQA scheme

This information can be invaluable. For example, the general standard of performance can emphasize the need for improvements in quality within a country, and monitor progress in improving between-laboratory agreement. EQA data can identify methods and techniques which yield better (or worse) performance in routine use than others, and thus encourage the adoption of more reliable procedures. The data obtained for each specimen should be compared with previous experience in the scheme, as this can provide a valuable real-time appraisal of the specimen distributed; if it shows problems, then individual laboratory performance should not be assessed on this material.

Overall, EQA schemes should not be for policing or licensing laboratories, but provide independent, objective data to act as an educational stimulus to improvement. Participation in EQASs should ideally be voluntary, certainly until confidence in the scheme's reliability has been established, though participation should also be seen as an integral part of professional standards in laboratory medicine.

5.2 Practical considerations

The practical factors associated with establishing EQA schemes were addressed by the World Health Organization (8) in the early 1980s. This publication also identified recommendations for EQAS design and operation, including the need for independence from any commercial manufacturing or marketing interests. The theoretical and practical considerations of EQA for immunodiagnostic investigations were comprehensively reviewed by Seth (19).

5.3 Scheme design

For an EQAS to be successful in stimulating improvement, participants must have confidence in the scientific validity of the scheme design as well as reliability of its operation, or they will not take action on information from the scheme. Experience with many schemes has indicated several essential design criteria (20):

(1) sufficient recent data, achieved through:
- frequent distributions
- rapid feedback of initial performance information after analysis

(2) effective communication of performance data, through:
- a cumulative scoring system
- structured, informative and intelligible reports

(3) an appropriate basis for assessment, including:
- stable, homogeneous specimens which behave like clinical specimens
- reliable and valid target values

5.3.1 Scoring systems

Scoring systems were introduced to make EQA information more comprehensible to participants. Comparison of results with targets in the form of overall or method-related means or histograms is difficult, especially with multiple specimens, and assessment 'by eye' of data accumulated over several distributions is yet more difficult. Anything other than a gross change in performance is therefore effectively impossible to detect.

Scoring makes such assessment much simpler through the cumulation of information in more readily comprehensible form. Cumulation may refer to a single analyte only, with use of results covering a period, to permit appraisal of performance relative to other participants at that time or to the individual laboratory's past performance; for the latter, the score must be independent of other participants' performance.

The most primitive systems give information in qualitative form, e.g. the 'pass/fail' criteria applied by licensing schemes, and thus give only the crudest reflection of performance. More sophisticated systems such as Variance Index (21) and BIAS/VAR (22) scoring yield quantitative information as a numeric score, retaining the potential for easy interpretation and being susceptible to graphical presentation. These have been of great assistance in enabling laboratories to recognize the existence of sub-optimal performance and in stimulating them to improve.

Most systems are based on an estimate of total error, but further refinement can also provide guidance on the type of contributory errors. Participants usually rely on precision information gained separately through the laboratory's IQC programme to assist in interpretation, with EQA providing estimates of bias and its consistency. A bias may be inconsistent due to poor within- or between-assay precision, but the potential presence of other contributory factors, such as non-linearity or other concentration-dependent bias, short- or long-term accuracy changes and specimen/method interactions, usually precludes such a simplistic interpretation. The BIAS and VAR parameters (22) provide an assessment of bias and its variability, both in percentage terms.

In the BIAS/VAR scoring system (used by UK NEQASs for hormone assay, and other schemes), the percentage bias of a participant's result relative to the target (see Section 5.3.3) is first calculated for each specimen. The cumulative bias (BIAS) is then the arithmetic mean of the Healy-trimmed bias values for all usable specimens (very high or very low analyte concentrations may be excluded, and individual laboratories may fail to return results) in the most recent six distributions (i.e. for up to 30 specimens with five-specimen distributions), usually requiring a minimum of six contributing biases. The cumulative variability of the bias (VAR) is the standard deviation of the bias values. Where results have been transformed logarithmically, the BIAS is the geometric mean, and the VAR is the corresponding geometric CV (22,23).

In the Variance Index system, the percentage biases are scaled using a Chosen Coefficient of Variation (CCV) reflecting the general standard of performance to yield a Bias Index Score (BIS) in a 'common currency':

$$\text{BIS} = (\text{percentage bias}) \times 100/\text{CCV}$$

The average BIS over a period (usually six distributions or ten scores, whichever is more) is calculated, yielding the Mean Running BIS (MRBIS), and their SD is the SDBIS. Interpretation of MRBIS and SDBIS is analogous to their unscaled counterparts, BIAS and VAR. Ignoring the sign of the percentage biases yields Mean Running Variance Index Score (MRVIS) as an index of total error. For guidance, an MRVIS below 100 is taken as a criterion of acceptable performance for many investigations, though each scheme has individual limits which are reviewed annually and dependent on the current standard of performance.

Table 1 shows an example of these calculations for total thyroxine for a participant laboratory in the UK NEQAS for Thyroid Hormones.

5.3.2 Report format

The most helpful scoring system will fail in its objective if the scores are not presented in such a way as to simplify the interpretation. Each participant has only limited resources (in terms of time, effort and ability) to devote to this interpretation, and experience suggests that those in most need of acting on EQA data devote or choose to devote the least.

An analyst receiving a report needs to make decisions on a series of questions, which are usually self-terminating when a negative answer is given:

- do I have a major overall problem?
- which analytes are contributing most to this?
- are these problems significant?
- what is the source of the errors in each case?

A well-designed combination of scoring system and report format can assist considerably in this process, and thus contribute to patient care not only through stimulation of improvement where this is indicated but also through removing the need for unnecessary investigation.

5.3.3 Target values

For participants' confidence in taking action based on EQA data, the targets against which performance is assessed are critical. The primary purpose of EQA is bias assessment, so the target should be accurate; similar considerations apply to value assignment for IQC materials. Ideally all target values should be validated fully, but this can present major difficulties and in practice their selection is based on a combination of prejudice, theory and practical considerations appropriate to the circumstances.

The main alternative approaches are:

- content derived from preparation (added analyte)
- definitive or reference method value (from analysis in one or more laboratories)
- reference laboratory value (from analysis in one or more laboratories)
- consensus value (mean or median derived from participants' results)

It is unusual that the target can be pre-determined by specimen composition, as this requires availability of analyte-free material, pure analyte for

Table 1. Calculation of scores (BIAS/VAR and Variance Index) for total thyroxine in UK NEQAS for Thyroid Hormones. Variance Index scores calculated using CCV of 10.0%

Distribution	Specimen	Laboratory result (nmol/litre)	Target (nmol/litre)	Bias (%)	BIS	VIS
195	966	47.0	50.89	–7.7	–76	76
	967	69.0	72.34	–4.6	–46	46
	968	117.0	123.76	–5.5	–55	55
	969	148.0	148.95	–0.6	–6	6
	970	148.0	152.19	–2.8	–28	28
196	971	160.0	172.71	–7.4	–74	74
	972	135.0	145.35	–7.1	–71	71
	973	152.0	166.87	–8.9	–89	89
	974	190.0	201.01	–5.5	–55	55
	975	119.0	124.48	–4.4	–44	44
197	976[a]	60.0	84.96	–29.4	–294	294
	977	123.0	122.98	0.0	0	0
	978	147.0	147.92	–0.6	–6	6
	979	147.0	150.42	–2.3	–23	23
	980	130.0	133.77	–2.8	–28	28
198	981	197.0	192.13	2.5	25	25
	982	166.0	164.28	1.1	10	10
	983	142.0	143.80	–1.3	–12	12
	984	121.0	120.31	0.6	6	6
	985	92.0	96.12	–4.3	–43	43
199	986	119.0	124.88	–4.7	–47	47
	987	192.0	202.25	–5.1	–51	51
	988	158.0	165.11	–4.3	–43	43
	989	138.0	144.99	–4.8	–48	48
	990	123.0	123.95	–0.8	–8	8
200	991	146.0	133.93	9.0	90	90
	992	152.0	152.06	0.0	0	0
	993	152.0	148.27	2.5	25	25
	994	128.0	123.73	3.5	34	34
	995	105.0	103.72	1.2	12	12
Total error		**MRVIS**				**45**
Bias		**BIAS**		**–2.7**		
		MRBIS			**–32**	
Variability of bias		**VAR**		**4.3**		
		SDBIS			**62**	

[a] Bias excluded from BIAS and VAR calculation.

addition, and no difference in behaviour between endogenous and added analyte. A single target value is obtained, as with definitive or reference methods, whereas the other two approaches can yield overall or method-dependent values.

Definitive or reference method values are theoretically the best option, but such methods are not available for all analytes, and the cost of obtaining them may be prohibitive; use of more than one laboratory is also desirable, to increase confidence in the resulting value (24). More seriously, however, the QC materials used in EQA and IQC often do not behave exactly as clinical specimens, giving different relationships between the values obtained by various methods; these matrix effects can then lead to incorrect conclusions being drawn about method accuracy.

The reference laboratory approach is only viable if the laboratories used have performance greatly superior to that of most participants, and in many cases this is not so or cannot be demonstrated. The dangers inherent in this approach are introducing errors due to the laboratories' own biases and giving these laboratories excessive confidence in their performance. Confidence in the mean is increased by use of several laboratories' values, with rejection of outliers.

There is no scientific reason for consensus values to be accurate, but practical experience shows that in most cases the mean of many participants' results is not only convenient (being essentially free, and available when required) but also sufficiently reliable as a target. Their validity must not, however, be assumed, but must be demonstrated through repeatability, recovery and comparison studies with other EQASs wherever possible. A sub-set of participants within a scheme, whose performance passes these validation tests, may be used to derive a target (8,19).

5.4 Competition in EQA

The competitive instinct is one factor motivating attempts to improve performance, and scoring facilitates competition in EQA, particularly when it includes a 'league table' presentation. Anything which improves comparability of results should be welcomed as benefiting patient care, but care must be exercised in exploiting this urge. Firstly it may encourage an attitude of improvement for improvement's sake, irrespective of clinical requirements, so that attaining good performance in EQASs becomes an end in itself.

More disturbingly, this may lead to a dissociation between the procedures used for EQA specimens and clinical specimens. Thus if assay replication and other favourable treatments (e.g. 25) are used, EQAS performance will no longer be an objective reflection of that normally attained in the laboratory. This is fundamentally dishonest, though such 'cheating' only deceives the participant laboratory into a false impression of the reliability of their assays.

For these reasons most scheme organizers try to avoid an excessive competitive element while still encouraging a healthy striving to emulate the performance of the best laboratories. For best effect, EQA specimens should be treated as similarly as possible to clinical specimens.

5.5 How to use EQA participation

EQA data should always be considered in conjunction with current IQC performance. Within the EQA report, interpretation should include appraisal relative to other users of the same system as well as compared to target values. Interpretation should be stepwise and to appropriate depth: for example, if performance is excellent, it is not necessary (and arguably an inappropriate use of professional time) to examine reports in great detail (20). A guide to interpreting UK NEQAS endocrinology reports is available in the participants' manual (26), but stages should include:

(1) check that the report refers to your laboratory (i.e. correct laboratory code);

(2) check that the results are those you obtained and reported;

(3) examine the percentage bias from target for each specimen and analyte in the distribution;

(4) compare current performance (e.g. BIAS and VAR) with other participants, and other users of the same method;

(5) examine performance scores for trends (e.g. BIAS and VAR against distribution graphs);

(6) compare the relationship of mean values for your method against the target;

(7) examine the relationship between your results and target values for the last six distributions.

5.6 Selection of an EQA scheme

Where alternative EQA schemes are available, laboratories need to make a choice between them, though in many cases participation in more than one scheme may be helpful in providing complementary information. In immunodiagnostic investigations particularly, EQA cost should be secondary to quality, and should be considered relative to reagent and equipment costs. Factors which should be considered include:

(1) scheme design validity, especially specimen appropriateness and concentration range, report and scoring quality (see Section 5.3 above);

(2) independence from manufacturing and marketing interests;

(3) scheme service and responsiveness, including turnaround and response to enquiries;

(4) reputation and previous experience; and

(5) cost.

6. Materials for IQC and EQA

Since fresh clinical specimens are not directly applicable, QC materials are required for most IQC and EQA procedures. Such materials must be fit for the intended purpose, and the main needs are for:

- homogeneity
- stability
- behaviour similar to clinical specimens
- availability in quantity
- affordable cost

In much of immunodiagnostics, specimens are of blood serum or plasma, and serum is the preferred basis for a QC material in view of its greater stability. Though animal serum is generally appropriate for both IQC and EQA of most clinical chemistry investigations, special problems arise where the analyte is species-specific (e.g. immunoglobulins) or the serum matrix is important in separation stages (e.g. immunoassays). For these reasons human-based QC materials are recommended for immunodiagnostic investigations, unless the appropriateness of animal-based materials has been proven.

6.1 Material sources and preparation

Most QC materials for immunodiagnostics are obtained from commercial sources. Such materials are normally produced in large quantity and stabilized, usually by lyophilization, and their behaviour may then not mimic that of patients' sera (lack of commutability, 'matrix effects'); problems are likely to be much more marked with automated immunoassay systems than with more manual procedures. It is essential in using commercial materials to be aware of the often limited range of suitability (sometimes to one system only), and balance the reassurance of knowing that a material has been validated for a procedure against the concerns of relying on QC material from the system manufacturer (calibration and control must not use the same material).

An alternative to use of commercial material is local preparation. It may be feasible to obtain serum from volunteer donors, but both ethical and safety aspects must be considered. Within the UK, all material used for IQC or EQA should have been tested for infective against to the same standard as blood products for therapeutic purposes (Department of Health circular HN(86)25); informed consent is required for testing at the individual donation stage for antibodies to HIV. Material which is analyte-free (or with low concentrations) for physiological reasons may be very helpful in assessing baseline security, interference or recovery (2,18,19).

Material with a range of concentrations should be available. Following preliminary analysis of the individual donations, it may be possible to prepare serum pools at appropriate levels. For many situations, however, it will be necessary to adjust analyte levels and prepare higher concentrations by spiking with pure analyte or an analyte-enriched preparation. The availability and purity of the material to be added will differ, and the effects of these factors must be considered. Selective depletion of analytes is possible in some cases, but the treatment will often change the material's properties (27).

Another approach to preparing concentrated or diluted materials uses a carefully mixed serum pool which is then frozen. On thawing the top layer will contain very low concentrations of all constituents. A volume of up to 25% of the total may be removed from the top, and the remaining concentrated serum then mixed carefully and thoroughly. Similarly, upper fractions may be used to yield a diluted serum. Full details are given in a WHO document (28), but it is important to note that use of this procedure can only increase or decrease the concentration of all serum components equally.

6.2 Material presentation

QC sera may be provided either in liquid state or freeze-dried (lyophilized), or could also be kept in frozen state in the user laboratory. Where the material is to be used strictly for internal quality control purposes the advantage of frozen liquid serum, distributed in suitable vials, is that accurate dispensing is not required and any error of reconstitution is obviated. Where sera have to be dispatched by post or other means it may be essential to use a lyophilized preparation.

6.2.1 Frozen liquid control sera

Frozen sera are intended primarily for internal control purposes within a laboratory. Storage should be at –30 °C or below; –20 °C is not adequate. Each working day one or more vials should be removed from the refrigerator, thawed under gentle conditions until all the ice has melted, after which it should be well mixed by gentle inversion.

6.2.2 Lyophilized sera

Lyophilized materials are intended for internal quality control purposes or use in external quality assessment schemes.

The volume of distilled water or reconstitution fluid used for reconstitution is critical. The exact volume should he stated on the label, and a special Grade 1 calibrated bulb pipette kept exclusively for this purpose. The cap should be removed slowly and carefully because particles of dried serum often adhere to the cap and must not be lost. The prescribed diluent or high quality distilled water is pipetted in, the cap replaced and the vial inverted three or four times and placed in a dark cupboard for five minutes, when it is taken out and again slowly inverted to aid solution. Vigorous shaking must be avoided. After 15 to 30 minutes solution should be complete and the material may be used.

It is important to remember that the solid material in the vial contributes to the volume of reconstituted serum, according to its protein content. A 10 ml vial reconstituted with 5 ml of water will therefore *not* have exactly double the usual concentrations.

6.2.3 Stabilized liquid serum

Ethylene glycol (ethanediol) can be an extremely useful and appropriate presentation for many investigations in clinical chemistry (28), but suitability for immunodiagnostic investigations has not yet been reported. Though azide is known to invalidate some enzyme assays and should not be used if the material is to be used for enzyme measurements, effects on enzyme-mediated immunoassays have been shown to be minimal. Azide (final concentration 15 mmol/litre) thus provides a practical and well-tried preservative for serum, as used in the UK NEQASs for Thyroid Hormones and for Steroid Hormones for many years (18).

6.3 Summary

Materials used for IQC and EQA must be appropriate for their purpose, which for immunodiagnostic investigations effectively requires a human base. In many cases, commercial materials are suitable, but divergence of behaviour from that with fresh human specimens must be considered. For IQC and for EQA, locally prepared liquid serum specimens should be considered, though the greater potential infection hazards must also be considered.

7. Discussion

7.1 Why use QA at all?

Quality assurance is often viewed as uninteresting, too expensive or too much trouble, so is it really necessary? Others see QA as something which a 'routine' laboratory undertakes but which is not applicable to 'research' situations.

These views represent an inappropriate response to a critical issue. Without appropriate and effective QA, what assurance is there that the results are fit for the intended purpose? This applies equally to providing a reliable clinical service to patients and to research (on which future patient treatment or scientific development may be based) from the immunodiagnostic laboratory. Costs of QA should always be measured against both the costs of carrying out the procedure (reagents, equipment, labour, etc.) and the costs (financial and clinical) of getting it wrong. Use of QA is therefore recognized as an integral part of competent professional practice, and not an expensive luxury.

7.2 What is good enough?

What are the limits of acceptability for immunodiagnostic investigations? Here the work undertaken on analytical goals in relation to biological variation provides a rational basis for assessment (29,30). Initial studies addressed primarily precision criteria, with the general guidance that analytical imprecision should not exceed half the within-individual variation, but the scope has been broadened to include assessment of factors such as bias, interference and exogenous analytes such as drugs. Continued development may be expected to provide a rational basis for future performance standards.

7.3 What reference intervals should be used?

Should reference intervals be universal or lab-specific? The objective should be for these to depend only on the population examined, being independent of the laboratory and the procedure used for assay. This has been pursued in the Nordic countries, with publication of common reference intervals for this relatively homogeneous population group. Extension is likely to demonstrate that ethnic origin is a more important determinant than geography (raising ethical considerations in obtaining this information with requests). For common reference intervals to be used, however, results must be analytically indistinguishable, and EQA will play a major role in this.

For results to be numerically as well as analytically comparable, however, a common unit set must be used. Here the molar SI system of units is preferred wherever possible; even where mass or unitage has to be used, the base unit for volume should be the litre (not ml or dl) and results should be reported in, for example, mU/litre rather than μU/ml.

7.4 Who has responsibility for quality?

Commercial products (kits, equipment and now automated systems) are used increasingly in immunodiagnostic laboratories. This has some benefits in passing responsibility for ensuring quality to the manufacturer as part of GMP and production QC. The drawback is that the laboratory has less ability to influence the quality of results obtained, and may have no possibility of doing so in some 'black box' analysers. If there are problems, laboratories tend to blame the supplier, and even where poor performance is demonstrated state that it is the manufacturer's fault and he has no alternative but to use the product. This is not, however, a professional attitude, and the laboratory staff must take responsibility for the results they report; if performance with a product is unacceptable, the only appropriate response is to stop using it and either use another product or refer specimens to another laboratory for assay.

7.5 Future prospects

Where is quality assurance for immunodiagnostic investigations going? With the increasing use of automated systems, and consequently greater emphasis on production QC and QA measures integral and internal to the systems, the emphasis is changing. Greater system stability means less frequent calibration and IQC, with fewer analytical failures. Continued vigilance is required, however, and EQA will fulfil another important role in assuring between-laboratory comparability and, through national or supra-national collation of data, post-market surveillance of performance of commercial systems. The basic principles of QA will always remain relevant despite these changes in analytical QC, and basic analytical QC will continue to be required for specialized and research applications of immunodiagnostics.

Acknowledgements

We are grateful to Professor Tom Whitehead and Professor Stephen Jeffcoate for their advice and support over many years; their publications (references 1 and 3) provide an invaluable source of practical and readable advice, and are highly recommended. We also thank our colleagues John Seth, Jonathan Middle, Jerry Snell, Callum Fraser and Colin Selby for helpful discussion, and all the staff of the Wolfson EQA Laboratory for their support.

References

1. Whitehead, T. P. (1977). *Quality control in clinical chemistry*. Wiley, New York.
2. Whitehead, T. P. (1976). *Principles of quality control (LAB/76.1)*. World Health Organization, Geneva.
3. Jeffcoate, S. L. (1981). *Efficiency and effectiveness in the endocrine laboratory*. Academic Press, London.
4. Fraser, C. G. (1986). *Interpretation of clinical chemistry laboratory data*. Blackwell, Oxford.
5. Seth, J. (1991). In *Principles and practice of immunoassay* (ed. C. P. Price and D. J. Newman), pp. 154–89. Macmillan, London.
6. Middle, J. G. (1995). In *Steroid analysis* (ed. H. L. J. Makin, D. B. Gower and D. N. Kirk), pp. 647–96. Blackie, London.
7. Selby, C. (1985). *Commun. Lab. Med.*, **1**, 8, 33 and 67.
8. World Health Organization (1981). *External quality assessment of health laboratories (EURO Reports & Studies 6)*. WHO Regional Office for Europe, Copenhagen.
9. Batstone, G. (1995) In *Proceedings of the UK NEQAS Endocrinology Meeting 1994* (ed. C. M. Sturgeon, J. Seth, J. G. Middle and S. P. Halloran), pp. 12–21. Association of Clinical Biochemists, London.
10. Freedman, D. B. (1995) In *Proceedings of the UK NEQAS Endocrinology Meeting 1994* (ed. C. M. Sturgeon, J. Seth, J. G. Middle, and S. P. Halloran), pp. 22–31. Association of Clinical Biochemists, London.
11. Audit Steering Committee (1991). *J. Clin. Pathol.*, **44**, 172.
12. Wheeler, M. J. (1995) In *Proceedings of the UK NEQAS Endocrinology Meeting 1994* (ed. C. M. Sturgeon, J. Seth, J. G. Middle and S. P. Halloran), pp. 163–70. Association of Clinical Biochemists, London.
13. Broughton, P. M. G., Bullock, D. G. and Cramb, R. (1989). *Brit. Med. J.*, **298**, 297.

14. Westgard, J. O., Barry, P. L., Hunt, M. R. and Groth, T. (1981). *Clin. Chem.*, **27**, 493.
15. Wheeler, D. J. (1983). *J. Quality Tech.*, **15**, 155.
16. Ekins, R. (1983). In *Immunoassays for clinical chemistry* (ed. W. M. Hunter and J. T. Corrie), 2nd edn, pp. 76–105. Churchill Livingstone, Edinburgh.
17. Spencer, C. A., Takeuchi, M., Kazarosyan, M., MacKenzie, F., Beckett, G. J. and Wilkinson, E. (1995). *Clin. Chem.*, **3**, 367.
18. MacKenzie, F. (1995) In *Proceedings of the UK NEQAS Endocrinology Meeting 1994* (ed. C. M. Sturgeon, J. Seth, J. G. Middle and S. P. Halloran), pp. 219–24. Association of Clinical Biochemists, London.
19. Seth, J. (1987). *J. Clin. Biochem. Nutr.*, **2**, 111.
20. Bullock, D. G. (1992). *Anal. Proc.*, **29**, 189.
21. Bullock, D. G. and Wilde, C. E. (1985). *Ann, Clin. Biochem.,* **22,** 273.
22. Bacon, R. R. A., Hunter, W. M. and McKenzie, I. (1983). In *Immunoassays for clinical chemistry* (ed. W. M. Hunter and J. T. Corrie), 2nd edn, pp. 669–79. Churchill Livingstone, Edinburgh.
23. Healy, M. J. R. (1979). *Clin. Chem.*, **25**, 675.
24. Stamm, D. (1982). *J. Clin. Chem. Clin. Biochem.*, **20**, 817.
25. Rowan, R. M., Laker, M. F. and Alberti, K. G. M. M. (1984). *Ann. Clin. Biochem.*, **21**, 64.
26. MacKenzie, F. (1997) In *Proceedings of the UK NEQAS Endocrinology Meeting 1996* (ed. C. M. Sturgeon, J. Seth, J. G. Middle and S. P. Halloran) , pp. 102–5. Association of Clinical Biochemists, London.
27. Jones, A., Ratcliffe, J. G. and Swift, A. D. (1986). *Commun. Lab. Med.*, **2**, 128.
28. Browning, D. M., Hill, P. G., Vazquez, R. and Olazabal, D. A. (1986). *Preparation of stabilized liquid quality control serum to be used in clinical chemistry (LAB/86.4).* World Health Organization, Geneva.
29. Subcommittee on Analytical Goals in Clinical Chemistry of the World Association of Societies of Pathology (1979). *Am. J. Clin. Pathol.*, **71**, 624.
30. Fraser, C. G. (1983). *Adv. Clin. Chem.*, **23**, 299.

List of suppliers

Abbott Diagnostics, Abbott House, Norden Road, Maidenhead, Berkshire SL6 4XF.

Advanced Magnetics Inc., Cambridge, Massachusetts, 02138, USA

Aldrich, The Old Brickyard, New Road, Gillingham, Dorset, UK

Amersham

Amersham International plc., Lincoln Place, Green End, Aylesbury, Buckinghamshire HP20 2TP, UK.

Amersham Corporation, 2636 South Clearbrook Drive, Arlington Heights, IL 60005, USA.

Anachem, Charles Street, Luton, Beds, LU2 0EB, UK

Anderman

Anderman and Co. Ltd., 145 London Road, Kingston-Upon-Thames, Surrey KT17 7NH, UK.

Bangs Laboratories Inc., Carmel, IN 46032–2823, USA

Baxter Travenol, Wallingford Road, Compton, Nr. Newbury, Berkshire RG16 0QW.

Bayer plc, Diagnostics Division, Bayer House, Strawberry Hill, Newbury, Berkshire RG14 1JA.

BDH, Poole, Dorset, UK

BDH/Merck Ltd., Hunter Boulevard, Magna Park, Lutterworth, Leicestershire

Beckman Instruments

Beckman Instruments UK Ltd., Oakley Court, Kingsmead Business Park, London Road, High Wycombe, Bucks HP11 1JU, UK.

Beckman Instruments Inc., PO Box 3100, 2500 Harbor Boulevard, Fullerton, CA 92634, USA.

Becton Dickinson

Becton Dickinson and Co., Between Towns Road, Cowley, Oxford OX4 3LY, UK.

Becton Dickinson and Co., 2 Bridgewater Lane, Lincoln Park, NJ 07035, USA.

Behring Diagnostics Inc., 151 University Avenue, Westwood, MA 02090 USA.

Bibby Sterilin, Stone, Staffordshire, UK

Bio

Bio 101 Inc., c/o Statech Scientific Ltd, 61–63 Dudley Street, Luton, Bedfordshire LU2 0HP, UK.

Bio 101 Inc., PO Box 2284, La Jolla, CA 92038–2284, USA.

Bio-Rad Laboratories

Bio-Rad Laboratories Ltd., Bio-Rad House, Maylands Avenue, Hemel Hempstead HP2 7TD, UK.

Bio-Stat Ltd, Bio-Stat House, Pepper Road, Stockport, Cheshire SK7 5BW.

Bio-Tek Kontron Instruments, 8 Marlin House, Croxley Business Park, Watford, Hertfordshire WD1 8YA.

Biogenesis Ltd, 7 New Fields, Stinsford Road, Poole, BH17 0NF, UK

BioMerieux, Grafton House, Basingstoke, Hampshire RG22 6HY.

Boehringer Mannheim GmbH, D–68298 Mannheim, Germany.

Boehringer–Mannheim GmbH, Postfach 31 01 20, Sandhofer Strasse 116, D-6800 Mannheim, Germany

Bronwes, Pincents Kiln Industrial Park, Calcot, Reading, RG31 7SB.

Byk Sangtek Diagnostica GmbH & Co. KG, von Hevesey-Strasse D–63128 Dietzenbach.

Canberra Packard, Pangbourne, Berkshire.

Chiron Diagnostics Ltd, Colchester Road, Halstead, Essex CO9 2DX.

CIS (UK) Limited, Dowding House, Wellington Road, High Wycombe, Buckinghamshire HP12 3PR.

Costar, Corning Costar Corp., Cambridge, MA02140, USA

DAKO Ltd, 16 Manor Courtyard, Hughenden Avenue, High Wycombe, Bucks. HP13 5RE, UK

DPL Division, EURO/DPC Limited, Glyn Rhonwy, Llanberis, Caernarfon, Gwynedd LL55 4EL.

Dynal, New Ferry, Wirral, Merseyside, UK

Dynatech, Billingshurst, W. Sussex, UK

Dynex Technologies, Daux Road, Billingshurst, West Sussex RH14 9SJ.

EG&G Wallac, 20 Vincent Avenue, Crownhill Business Centre, Crown Hill, Milton Keynes MK8 0AB.

Eppendorf pipettes and other laboratory equipment and reagents:

Europa Bioproducts Ltd, Europa House, 15-17 North Street, Wicken, Ely, Cambridge CB7 5XW.

Fluka, Gillingham, Dorset, UK

Greiner, GreinerLabrotechnik Ltd, Stonehouse, Gloucestershire, UK

Hamilton GB Ltd, Unit 2, Lyne Riggs Estate, Lancaster Road, Carnforth, Lancashire LA5 9EA.

Hycel Groupe Lisabio, 12 avenue Charles-de-Gaulle, 91420 Morangis, France.

ICN, ICN Biomedicals Inc., Costa Mesa, CA92626, USA

JBL Scientific Inc., 277 Granada Drive, San Luis Obispo, California 93401-7396 USA

Labsystems, P.O. Box 8, IN–00881 Helsinki, Finland.

Life Technologies Ltd, PO Box 35, Trident House, Renfrew Road, Paisley PA3 4EF

Molecular Devices Corporation, 3180 Porter Drive, Palo Alto, CA 94304, USA

Molecular Devices Ltd, Unit 4, Charlwood Court, Country Oak Way, Crawley, West Sussex RH11 7XA.

Murex Diagnostics Ltd, Central Road, Temple Hill, Dartford, Kent, DA1 5LR, UK

NBL Gene Science, Cramlington, Northumberland, UK

Nichols Institute Diagnostics B.V., Kerkenbos 1051, 6546 BB Nijmegen, The Netherlands.

Novo Nordisk. Novo Allé, DK 2880 Bagsværd, Denmark

Nunc A/S, Postbox 280, DK-4000 Roskilde, Denmark

Nunc, Kamstrup, DK 4000, Roskilde, Denmark

Organon Teknica nv, Veedijk 58, 2300 Turnhout, Belgium.

Ortho Clinical Diagnostics Ltd, Mandeville House, 62 The Broadway, Amersham, Buckinghamshire HP7 0HJ.

Perkin Elmer (PE) Applied Biosystems, Birchwood Science Park North, Warrington, Cheshire WA3 7PB.

Pharmacia Biosystems

Pharmacia Biosystems Ltd. (Biotechnology Division), Davy Avenue, Knowlhill, Milton Keynes MK5 8PH, UK.

Pharmacia LKB Biotechnology AB, Björngatan 30, S-75182 Uppsala, Sweden.

Pharmacia, St Albans, Hertfordshire, UK

Pierce and Warriner (UK) Ltd, 44 Upper Northgate Street, Chester, CH1 4EF, UK

Pierce, Pierce and Warriner, Chester, UK

3747 N. Meridian Road, PO Box 117, Rockford, IL 61105, USA

Promega

Promega Ltd., Delta House, Enterprise Road, Chilworth Research Centre, Southampton, UK.

Promega Corporation, 2800 Woods Hollow Road, Madison, WI 53711–5399, USA.

Qiagen

Qiagen Inc., c/o Hybaid, 111–113 Waldegrave Road, Teddington, Middlesex, TW11 8LL, UK.

Qiagen Inc., 9259 Eton Avenue, Chatsworth, CA 91311, USA.

R & D Systems Europe, Ltd, 4–10 The Quadrant, Barton Lane, Abingdon, OX14 3YS

Radim S.p.A. Via del Mare, 125–00040 POMEZIA (ROMA)–ITALIA – UK distributor – TRIO Diagnostics, Chessington Park, Common Road, Dunnington, York YO1 5SE.

Reactifs IBF–Pharmindustrie, 92115 Clichy, France

Roche Diagnostics, F. Hoffman–La Roche Ltd, Diagnostics division, 4070 Basal, Switzerland.

Rosys Anthos, Anthos Labtech Instruments Ges.m.b.H., Lagerhaussetrasse 507, A–5071 Wals/Salzburg, Austria.

Labtech International Limited, 1 Acorn House, The Broyle, Ringmes, East Sussex, BN8 5NW.

Sanofi Pasteur, 3 Boulevard Raymond Poincare – BP3, 92430 Marnes-La-Coquette, France.

Sarstedt, Leicester, Leicestershire, UK

Schleicher and Schuell

Schleicher and Schuell Inc., Keene, NH 03431A, USA.

Schleicher and Schuell Inc., D-3354 Dassel, Germany.

Schleicher and Schuell Inc., c/o Andermann and Company Ltd.

Scipac, Sandwich, Kent, UK

Serono Diagnostics Ltd, Shirley Avenue, Vale Road, Windsor, Royal Berkshire SL4 5LF.

Shandon Scientific Ltd., Chadwick Road, Astmoor, Runcorn, Cheshire WA7 1PR, UK.

Sigma Chemical Company

Sigma Chemical Company (UK), Fancy Road, Poole, Dorset BH17 7NH, UK.

Sigma Chemical Company, 3050 Spruce Street, P.O. Box 14508, St. Louis, MO 63178–9916.

Sorin Biomedica Belgium S.A./N.V., Rue de la Grenouillette 2F, Waterranonkelstraat 2F, 1130 Bruxelles/Brussel.

Sorvall DuPont Company, Biotechnology Division, P.O. Box 80022, Wilmington, DE 19880–0022, USA.

Stratagene

Stratagene Ltd., Unit 140, Cambridge Innovation Centre, Milton Road, Cambridge CB4 4FG, UK.

Strategene Inc., 11011 North Torrey Pines Road, La Jolla, CA 92037, USA.

Tecan UK Ltd., 18 High Street, Goring on Thames, Reading RG8 9AR

Tosoh Corporation Headquarters, Scientific Instrument Division, 1–11–39, Akasaka, Minato-Ku, Tokyo 107, Japan.

Eurogenics UK Ltd, Unit 5, Kingsway Business Park, Oldfield Road, Hampton TW12 2HD

United States Biochemical, P.O. Box 22400, Cleveland, OH 44122, USA.

Vector Laboratories Inc., 30 Ingold Road, Burlingame, CA 94010, USA, 3 Accent Park, Bakewell Road, Orton Southgate, Peterborough PE2 6XS, UK.

Wallac
Wallac Oy, PO Box 10, FIN-20101 Turku, Finland. E.G. & G. Wallac UK Ltd, 20 Vincent Avenue, Crownhill Business Centre, Crownhill, Milton Keynes, MK8 0AB

Wellcome Reagents, Langley Court, Beckenham, Kent BR3 3BS, UK.

Whatman, Maidstone, Kent, UK

Zenyx Scientific Ltd, Broad Oak Business Centre, Ashburton Road West, Trafford Park, Manchester M17 1RW.

Zinsser Analytic (UK) Ltd, Howarth Road, Stafferton Way, Maidenhead, Berkshire SL6 1AP.

Index

Index